Airbus A320 Abnormal Operation

Aviation Bookstore
aviation step by step

NOTE: *information not valid for real flights. Concepts explained as examples for your study and entertainment. For real flights see the original manuals of the manufacturer AIRBUS.*

Published by Aviation Bookstore LLC ®

Aviation Bookstore LLC ® is a registered trademark of Aviation Bookstore LLC.
Independent publishing brand, not affiliated with Airbus SAS, Boeing Company, or any other aeronautical institution.

The content of this book is developed for educational and professional training purposes.
Original text authored by humans.

Preface

In this edition, we will know all the abnormal operation of one of the most sold and flown commercial aircraft in the commercial aviation. We will know everything about the fabulous Airbus 320. We will learn the abnormal operation of the main systems of the airplane. How each of them works and how they are operated by the pilots from the control panels in the cockpit. A practical guide, didactic and entertaining for any professional who is about to start flying A320 or for any professional who wants to expand their frontiers of knowledge! This edition of the most prestigious collection in Latin America promises to mark the difference in the way of learning the systems of an airplane.

Capt. Facundo Conforti

NOTE: *information not valid for real flights. Concepts explained as examples for your study and entertainment. For real flights, see the original manuals of the manufacturer AIRBUS.*

Aviation Bookstore
aviation step by step

Main

Chapter 1 - Abnormal operations I

Auxiliar power unit 09
Pneumatic system 16
Electrical system 24
Air conditioning system 40

Chapter 2 - Abnormal operations II

Doors system 51
Fuel system 53
Hydraulic system 64
Navigation system 77

Chapter 3 - Abnormal operations III

Power plant system 89
Ice & Rain protection system 101
Fire protection system 105
Communication system 112

Chapter 4 - Abnormal operations IV

Landing gear system	119
EIS ECAM system	125
Ventilation system	129
Pressurization system	135
FLight control system	141

Chapter 1

Abnormal operations I

Aviation Bookstore

aviation step by step

NOTE: *information not valid for real flights. Concepts explained as examples for your study and entertainment. For real flights see the original manuals of the manufacturer AIRBUS.*

Aviation Bookstore
aviation step by step

APU abnormal operation

When the APU fail, the aircraft lost additional electrical and pneumatic power. In this case, pilots must to take action to solve the fails as soon as possible. Usually the APU fails when starting. When the system detects the fail, trigger the “Master Caution” alarm to alert pilots. It’s not an emergency situation but require actions by pilots.

This is a demonstration of an APU auto shutdown as a result of high EGT. You have left the runway, and you want to start the APU. Let’s start the APU with the MASTER SW pb and check what happen!

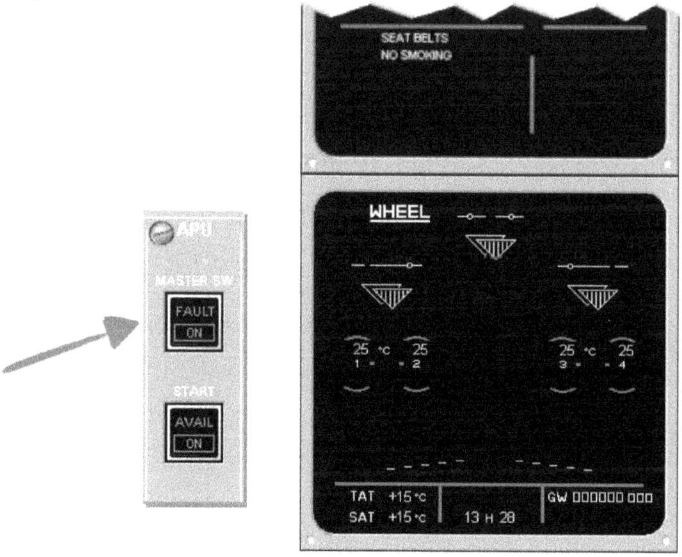

Now select the APU START pb sw to switch it ON.

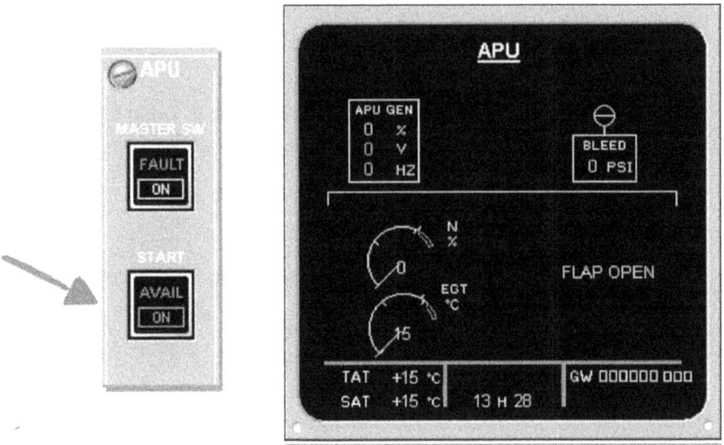

The procedure is finishing. The APU system is initiating normally Suddenly the MC alarm is triggered!

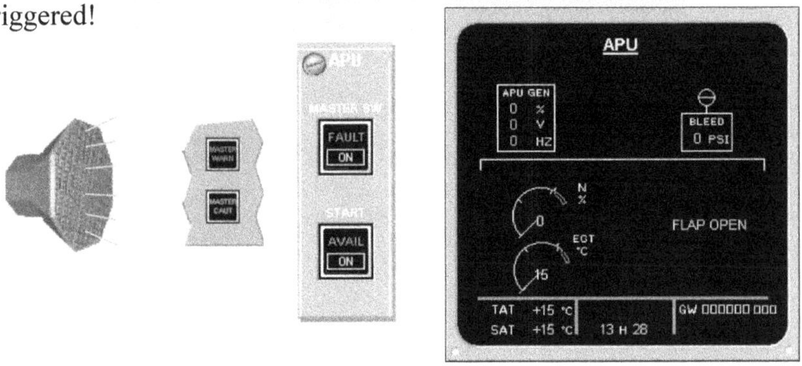

The system has detected a fail during starting cycle and the MC alarm has been triggered to alert pilots. Let's check the ECAM.

On the E/WD, you can read the title of the failure. The ECAM APU page is automatically displayed.

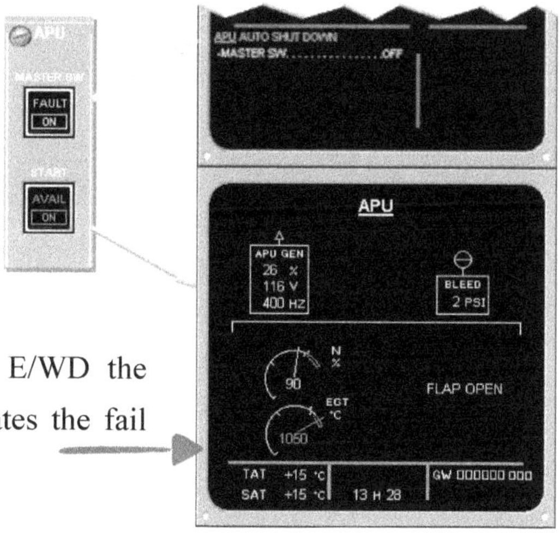

As we can see, on the E/WD the message on amber indicates the fail and the action required.

The ECAM caution "APU AUTO SHUT DOWN" appears amber to indicate that the APU has initiated an automatic shutdown. This is because the EGT has exceeded the maximun limit. Notice the red EGT indication on the ECAM APU page.

On the APU control panel, the FAULT light is illuminated on the MASTER Switch :

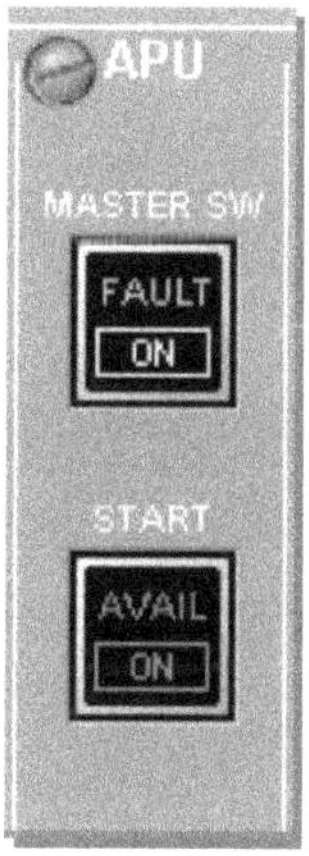

•this provides a cross check of the ECAM,

•and locates the pushbutton to be activated in the procedure.

As you can see, the APU has shutdown by itself. An automatic shutdown can also occur for other reasons, for example over speed, slow start, low oil pressure, etc. Now the Start button is not available.

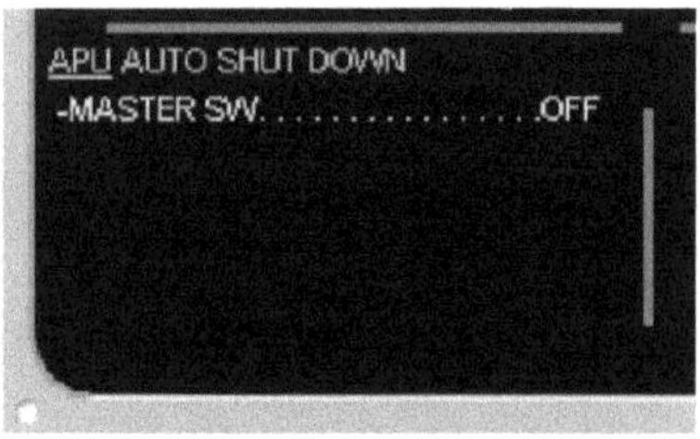

The PF will ask you to perform the ECAM actions.

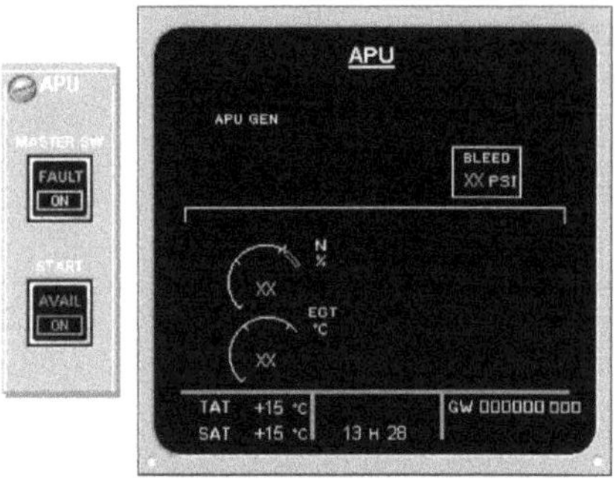

Observe that the blue ON light in the MASTER SW pb, and the FLAP OPEN indication on the ECAM have disappeared indicating that the APU flap is fully closed.

The abnormal procedure is complete, the APU system is analyzed.

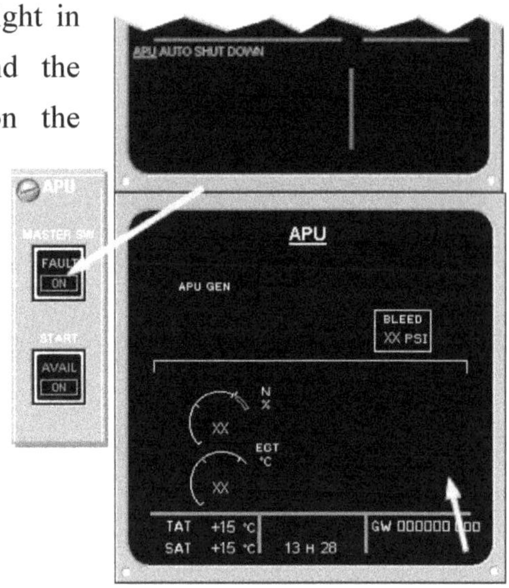

After review and confirmation from the PF, check STATUS. The STATUS page is displayed. The inoperative system is the APU and the FAULT message in amber is shown on the master sw.

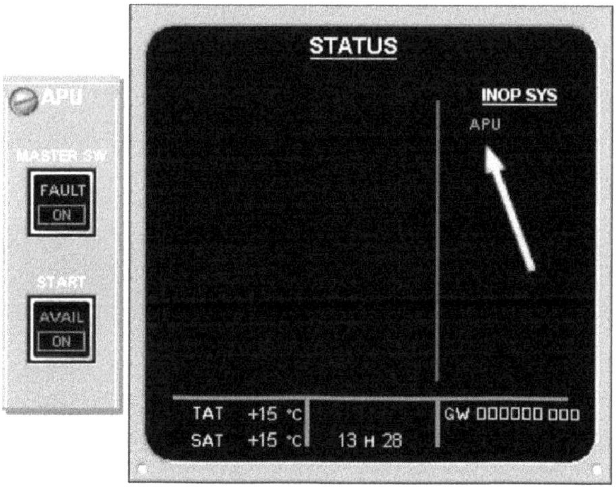

ECAM action completed. The STS reminder is displayed on the E/WD. Note : automatic shutdown protection is available throughout the flight.

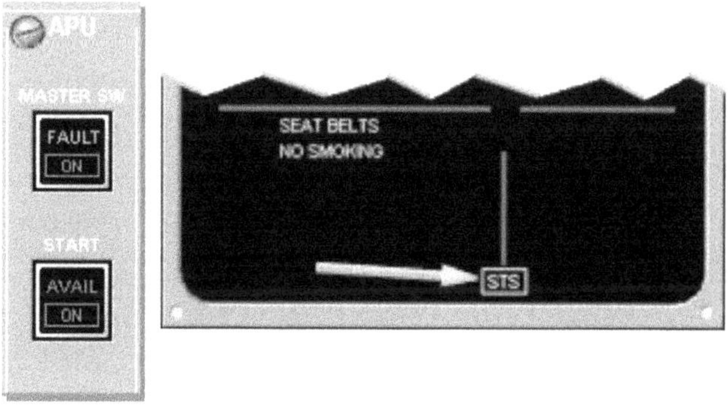

Let's now look at a situation which is similar to an APU AUTO SHUTDOWN but more critical, the APU EMERGENCY SHUTDOWN.

If, in a same situation, an APU fire is detected on the ground, an APU EMER SHUTDOWN occurs and the APU fire extinguisher bottle is automatically discharged (both without crew action). This will not happen in flight.

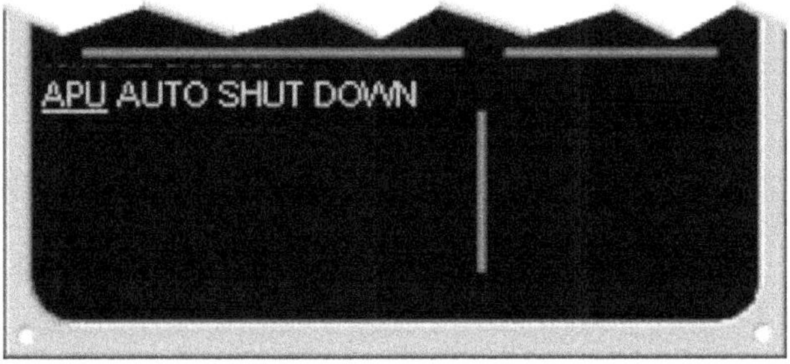

An APU emergency shutdown also occurs:

* When the APU SHUT OFF pb on the external power panel is pressed,

* Or when the APU FIRE pb on the FIRE panel is pressed.

In both cases, the fire extinguisher bottle is not automatically discharged.

On the ground :

- An APU AUTO SHUT DOWN occurs for a reason other than a fire or when APU parameters are exceeded.

- If a fire is detected, an APU EMER SHUT DOWN occurs with automatic extinguishing.

In flight :

- An APU AUTO SHUT DOWN occurs for the same reasons.

- If there is a fire, the APU EMER SHUT DOWN procedure is displayed at the end of the APU FIRE procedure. There is no automatic extinguishing.

Now let's look at the remaining indications that you could see on the ECAM page. The LOW OIL LEVEL advisory pulses in green if the APU oil quantity approaches its minimum value.

Note: minimum quantity allows normal APU operation for an additional 60 hours.

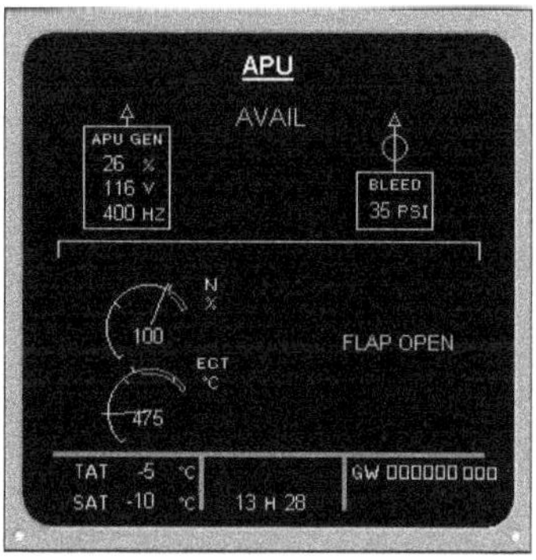

When starting the APU, the FUEL LO PR indication appears amber when an APU fuel low pressure is detected. Note : Even though this message is in amber, it is an advisory.

Pneumatic Abnormal Operation

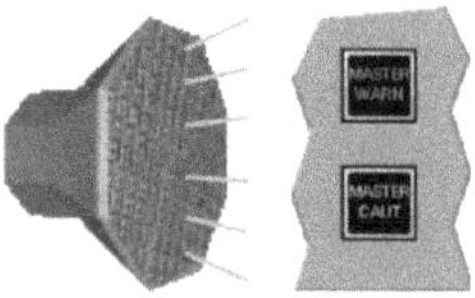

When the Pneumatic system fails, one or both the engine bleeds are affected. We will demonstrate an ENG 1 BLEED FAULT due to an overheat condition. Let's assume you are pilot non-flying. You are in cruise flight and everything is normal and suddenly the MC is triggered.

The ENG 1 BLEED system has failed. On the BLEED system page, which has been called automatically, the bleed air temperature is displayed amber because it is abnormally high. This is the reason for the failure.

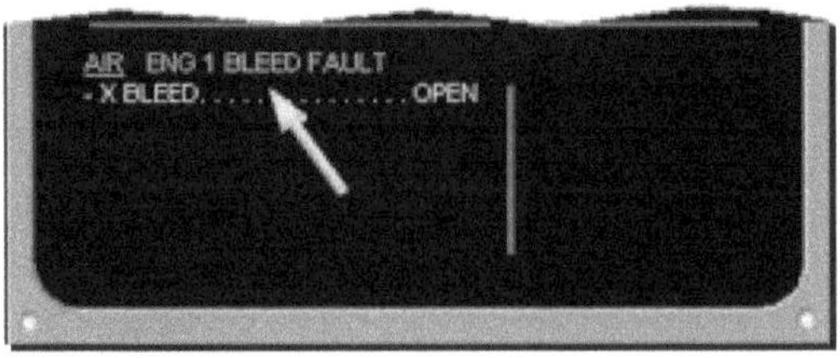

The amber FAULT light on the ENG 1 BLEED pb sw confirms the failure.

The BMC detects the overheat and automatically closes the ENG 1 BLEED valve to protect the system. With the valve closed, the BLEED pressure falls to 0 and the temperature decreases. The ENG 1 BLEED system is no longer available.

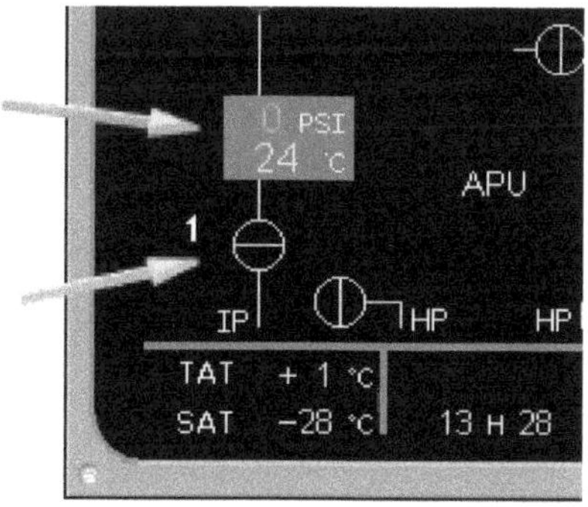

Notice that the ENG 1 BLEED pb sw FAULT light remains illuminated even though the temperature is now within limits. This is because the BMC considers the ENG 1 BLEED system faulty and keeps the ENG 1 BLEED valve closed for the rest of the flight. The PF will ask you to perform the ECAM actions.

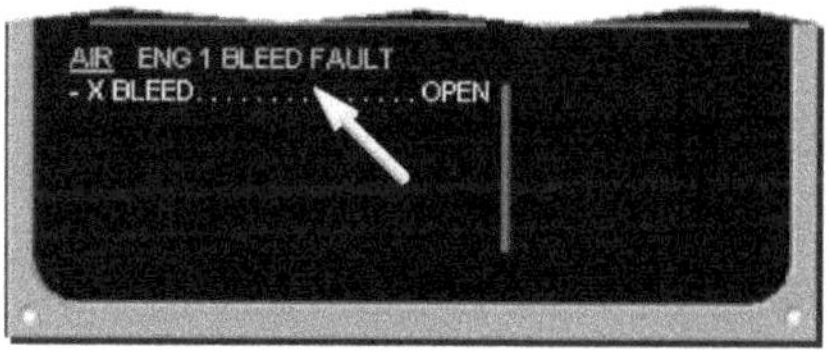

The PF set X-BLEED to OPEN from the OVERHEAD PANEL. After that, observe the ECAM.

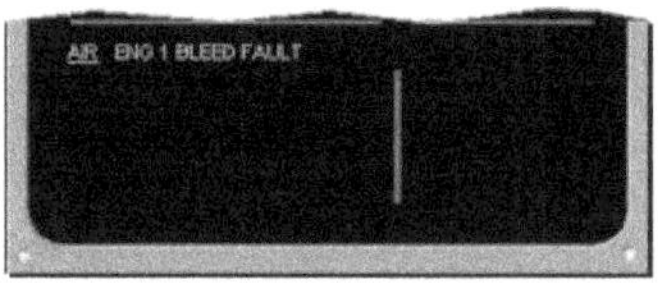

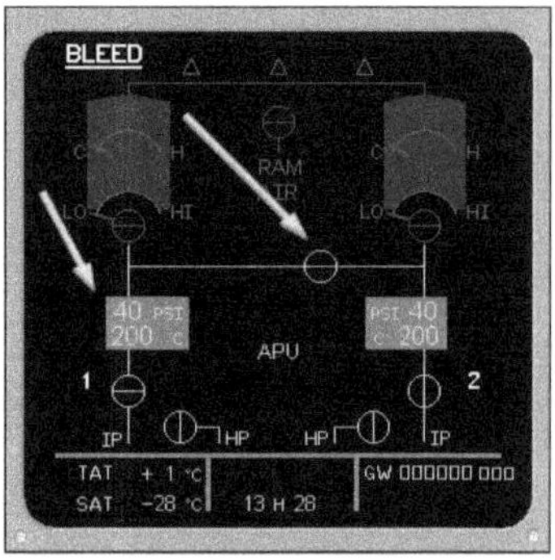

The X Bleed valve is displayed amber during transit and in-line green when fully open. The pressure and temperature values for the left side of the system are back to normal because now both sides are being supplied by engine 2.And the message on the E/WD is still in amber.

After ECAM ACTIONS, check status. If you should use wing anti ice, you will have to switch off one air conditioning pack. We will deal with this fact later.

In the inoperative systems column, ENG 1 BLEED is listed.

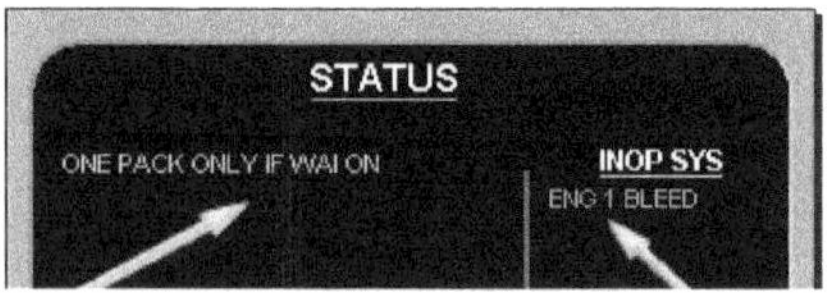

Now we will see what happens if you have to use wing anti ice with the ENG 1 bleed still inoperative. Let's assume, you encounter icing conditions during the remaining part of your flight and you decide to switch on your wing anti ice. After switching Wing Anti Ice ON, the MC alarm is triggered.

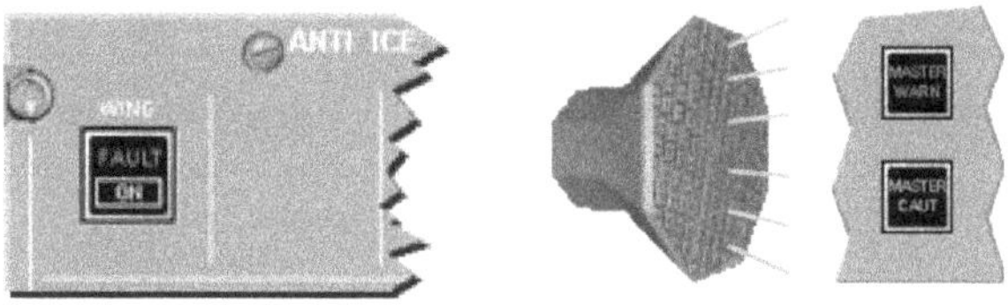

The ENG 1 BLEED FAULT procedure is automatically recalled on the E/WD and the ECAM procedure requires that you switch pack 1 off. This is because one ENG BLEED system is not sufficient to supply wing anti ice and both air conditioning packs at the same time. **Note:** For more information about the packs, refer to the AIR CONDITIONING chapter.

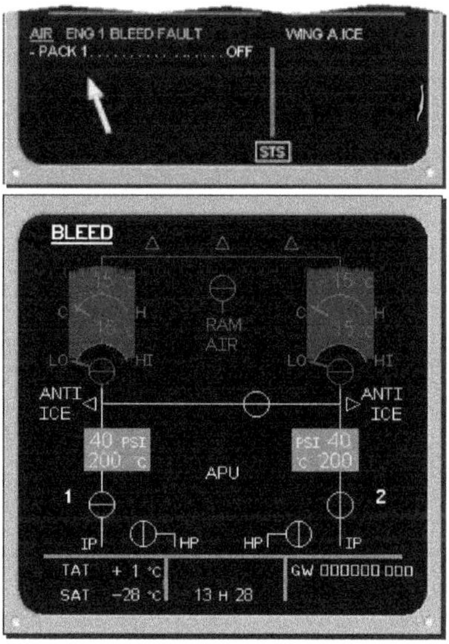

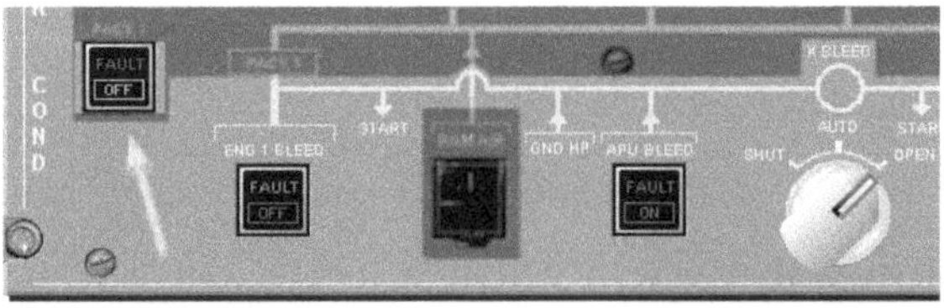

On the STATUS page, PACK 1 has been added to the inoperative systems column. After review and confirmation by the PF..

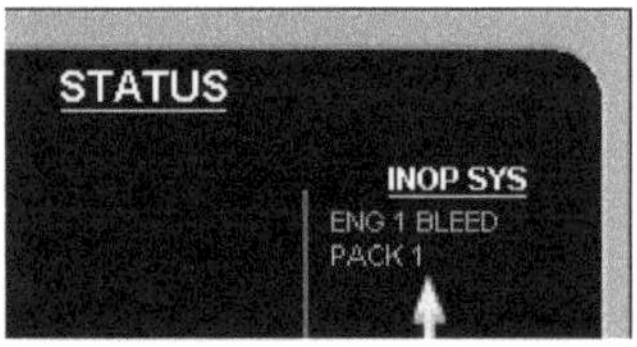

On the AIR COND panel, the indications from the previous failure remain unchanged.

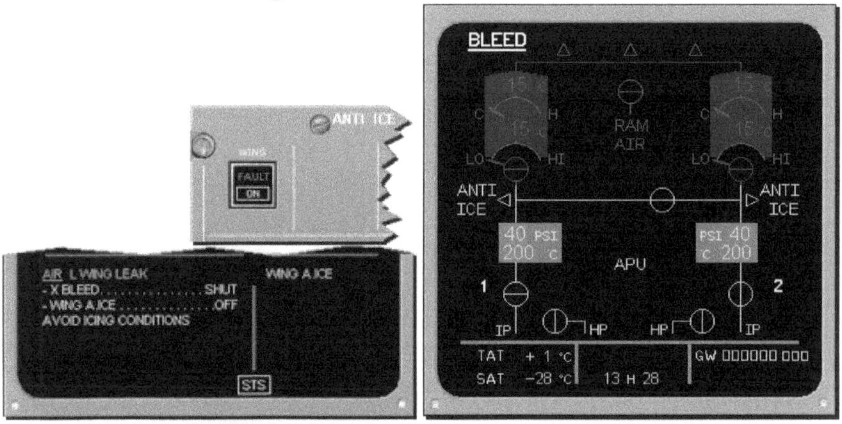

The title of the failure indicates that a bleed leakage in the left wing has been detected. The left side of the bleed system has to be shut down in order to prevent any damage from leaking hot air. The first line of the ECAM procedure requires you to switch the X BLEED selector to the SHUT position. With the selector in AUTO position, the X BLEED valve would have been closed automatically but remember, you have opened it manually in the previous procedure. The PF will ask you to perform the ECAM actions.

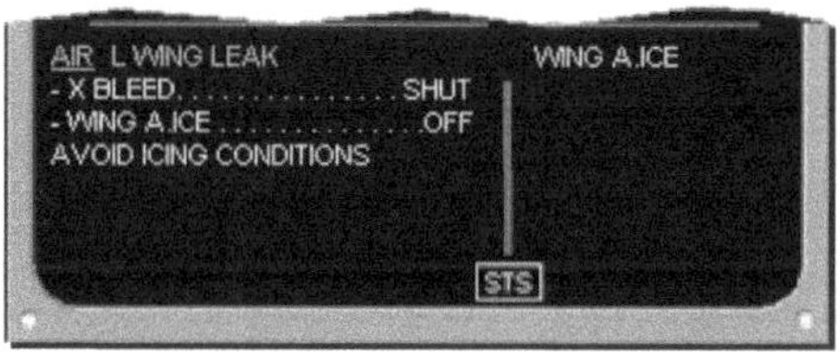

X-BLEED SHUT!

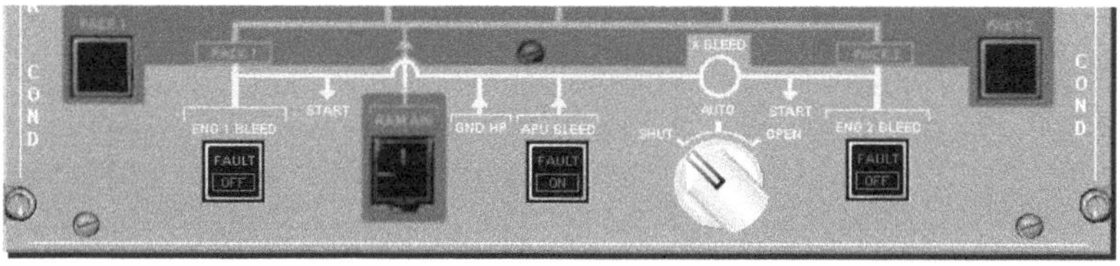

Observe the indications :

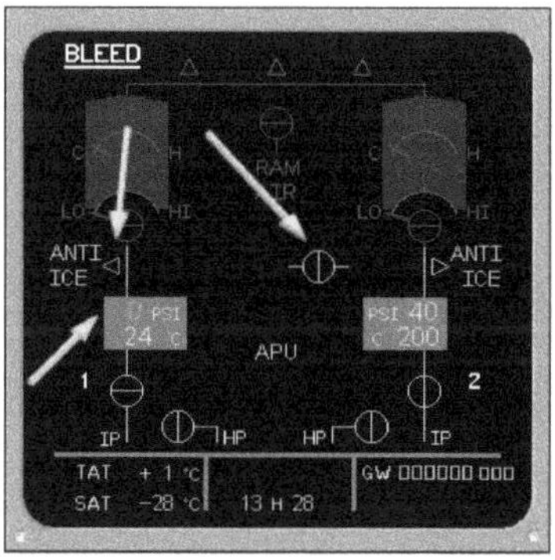

- The X BLEED valve closes.

- The pressure of BLEED system 1 returns to 0.

- Since there is no hot air

supplied, the left wing anti ice valve indication becomes amber. (the wing anti ice indications will be explained further in a later module)

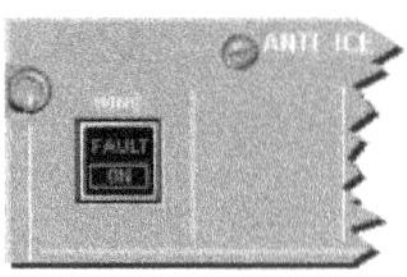

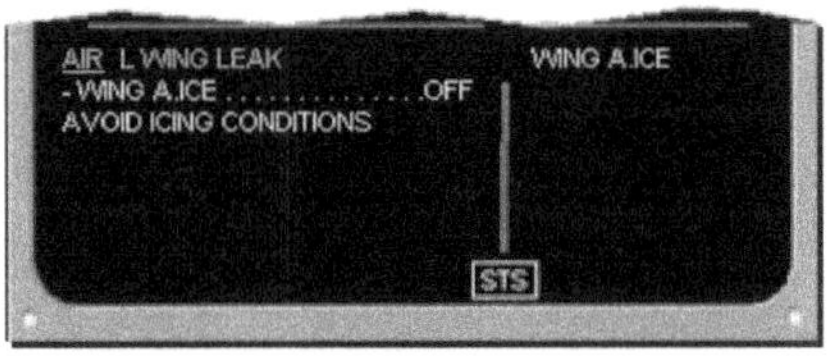

The wing anti ice was switched off because you do not want to de-ice one wing only. The ECAM advises you to avoid icing conditions for the remaining part of the flight.

The STATUS page also informs you that you have to avoid icing conditions. The WING ANTI ICE has been added to the inoperative systems column. The ENG 1 BLEED and PACK 1 remain from the previous failure.

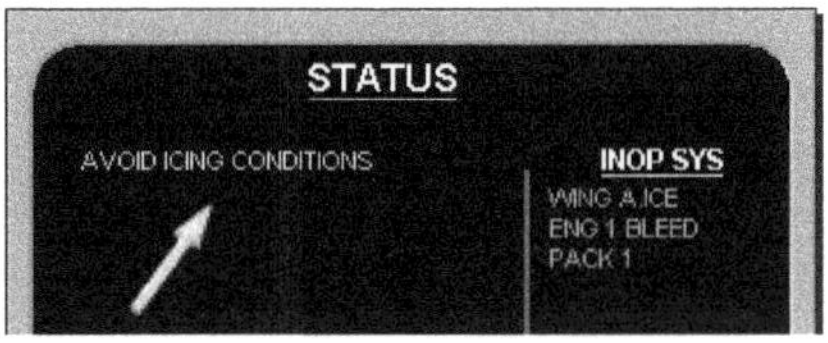

We will now briefly look at several abnormals without performing the procedures to see the remaining indications of the pneumatic system. In normal operation, when you use APU BLEED, the X BLEED valve automatically opens. If it should fail to open, there is an ECAM caution and the X BLEED valve is displayed cross-line amber to indicate disagreement.

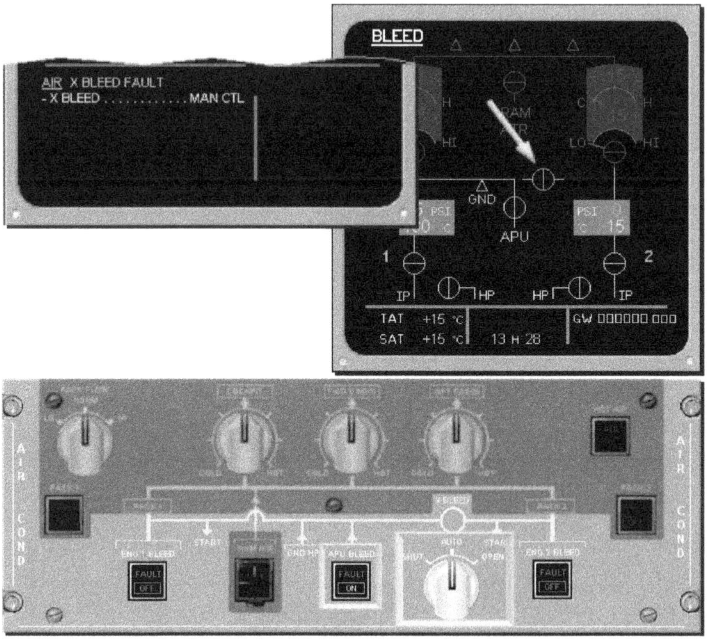

Conversely, if the X BLEED valve should fail to close in automatic mode, i.g. after the APU BLEED has been switched off, it is displayed in-line amber accompanied by an ECAM caution.

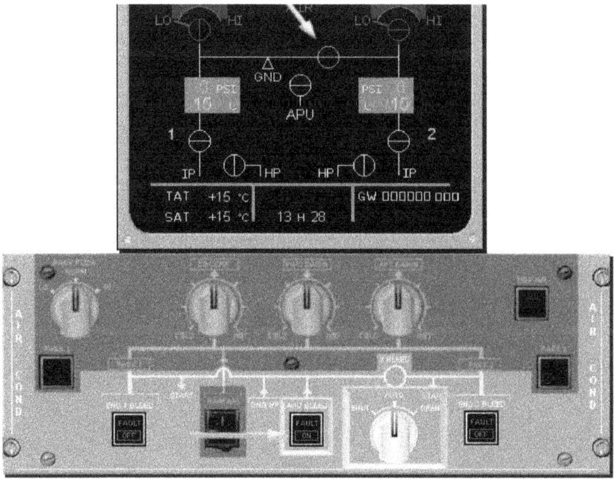

Accordingly, the engine bleed valves and the HP valves are displayed amber, if their position disagrees with the normal position.

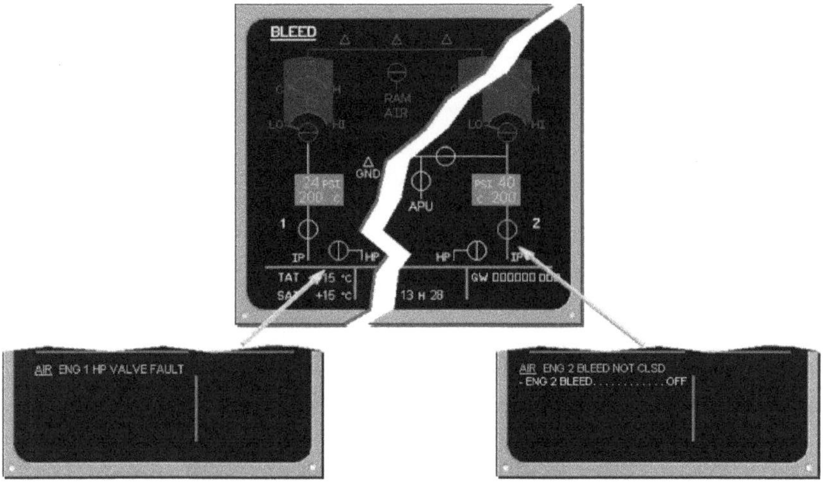

Electrical Abnormal Operation

Electrical fails are one of the most dangerous situation in the A320 and it requires some special attention by pilots. We will start with an IDG OIL LOW PRESSURE fault. Let's assume, you are the PNF. You are in flight and everything is normal, suddenly the MC is triggered!

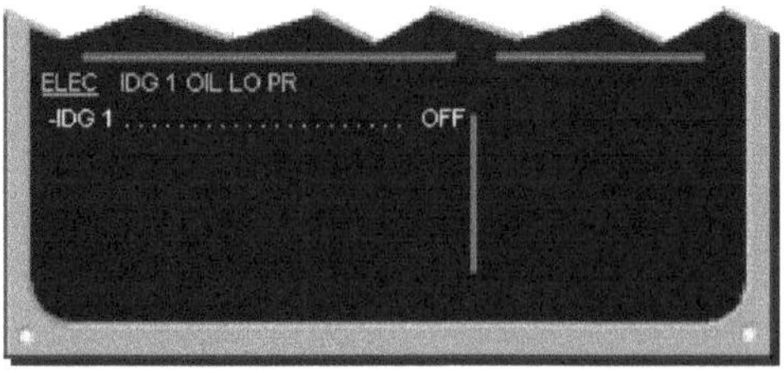

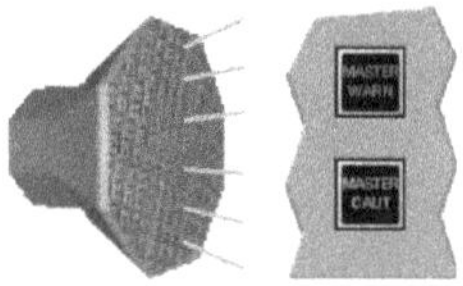

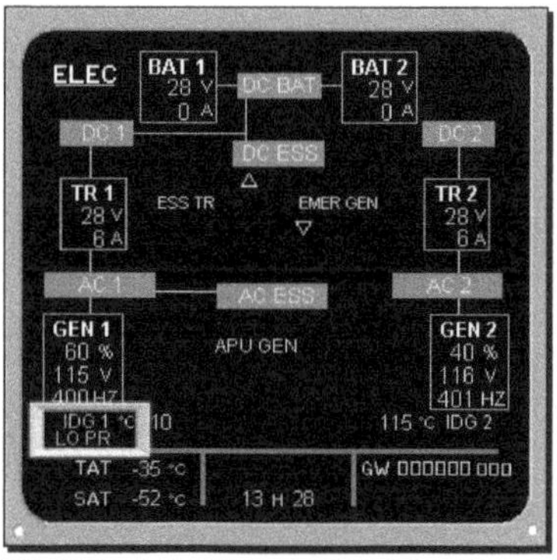

On the E/WD, read the title of the failure. Observe the IDG LO PR in amber on the ECAM ELEC page.

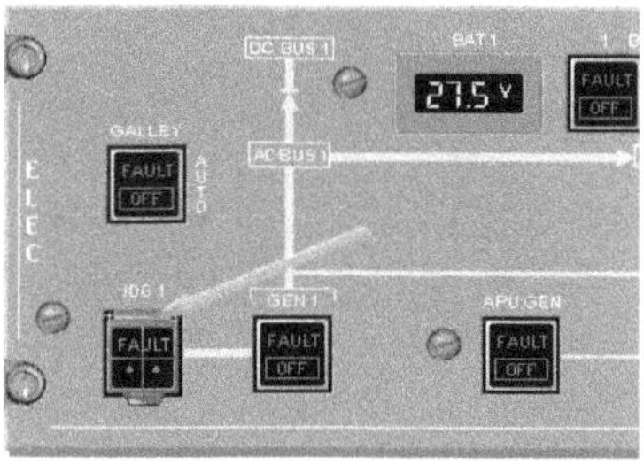

Observe IDG 1 FAULT light illuminated on the ELEC panel.

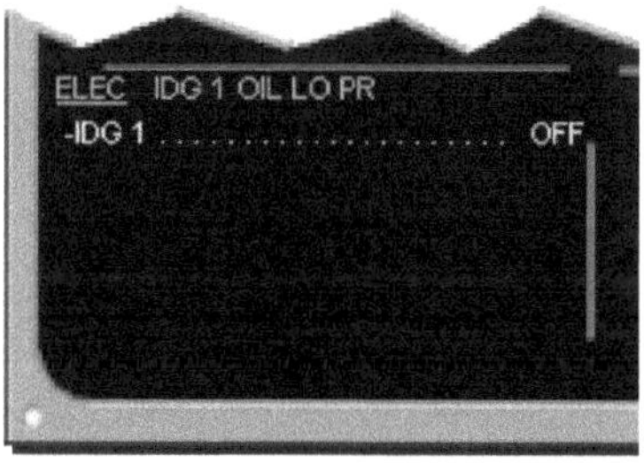

These indications mean that the oil pressure in IDG 1 is not sufficient. We can see from the blue action line that we are required to disconnect the IDG from the engine. The PF will ask you to perform the ECAM actions.

The pb sw guard is up, you can now disconnect IDG 1.

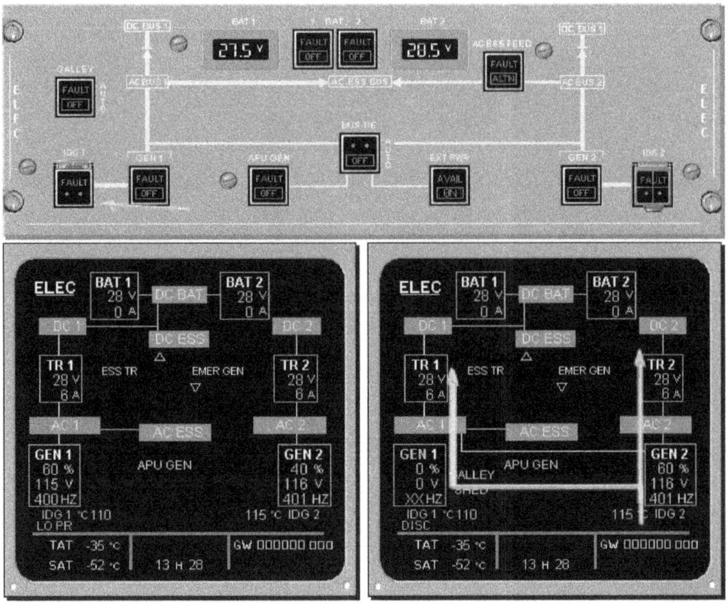

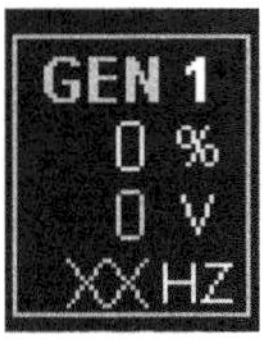

As you can see, generator 2 has automatically taken over and it is now supplying the entire electrical network. On the ECAM ELEC page, an amber DISC has replaced the previous LO PR meaning that IDG 1 is now disconnected. The IDG number is in white to indicate that engine 1 is still running. Let's look at GEN 1 indications for a moment.

- GEN in amber means generator 1 has abnormal parameters
- 1 in white means engine 1 is still running.
- 0% load in green means that the load is still within the normal range.
- 0 volts in amber means that the voltage is outside the normal range.
- Amber "X"s means that there is no frequency sensed.

FAULT amber illuminated in the GEN 1 pb sw means GEN 1 is not supplying the system.

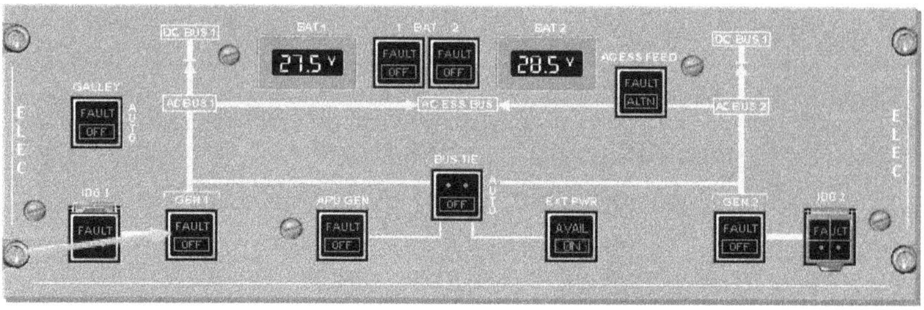

Alongside the GEN 1 box, a white GALLEY SHED indication has appeared meaning that the system has automatically shed galley equipment to decrease the load on the remaining generator.

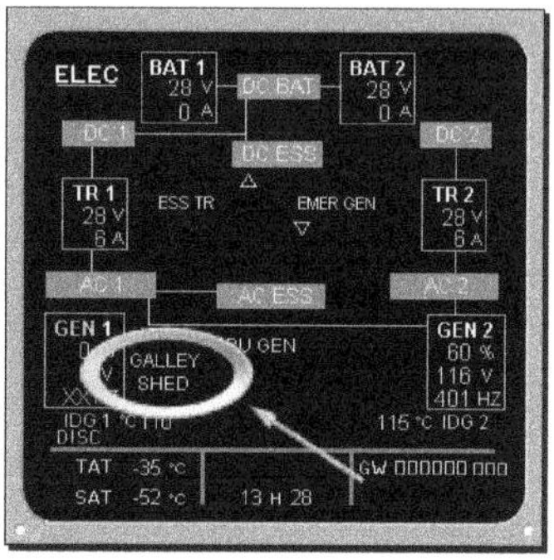

The STATUS page is displayed for review. A green message advises us that the APU generator is available up to FL 390, should we want to use the APU as a backup. You can see, in the INOP SYS column, that GEN 1 and the MAIN GALLEY are now inoperative.

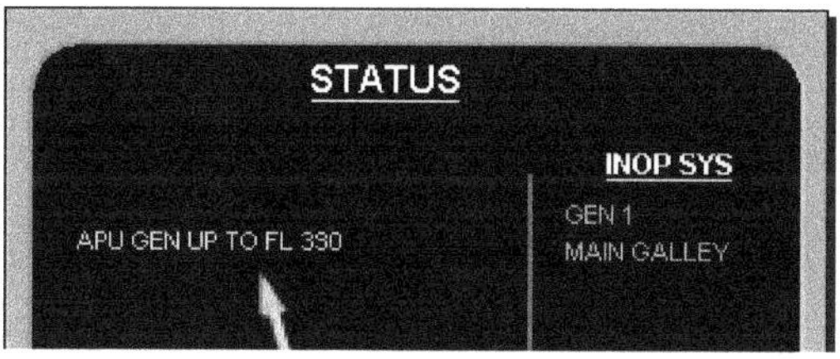

You have studied an IDG oil low pressure. We will now see a GEN 2 OVERLOAD fault. Let's assume we are in flight with no previous faults. Suddenly the MC alarm is triggered. On the E/WD, read the title of the failure.

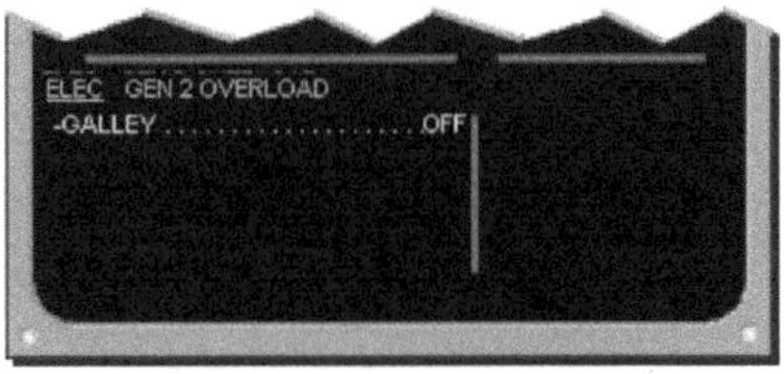

Observe:

GEN and load indications in amber on the ECAM ELEC page indicating that the load is abnormal.

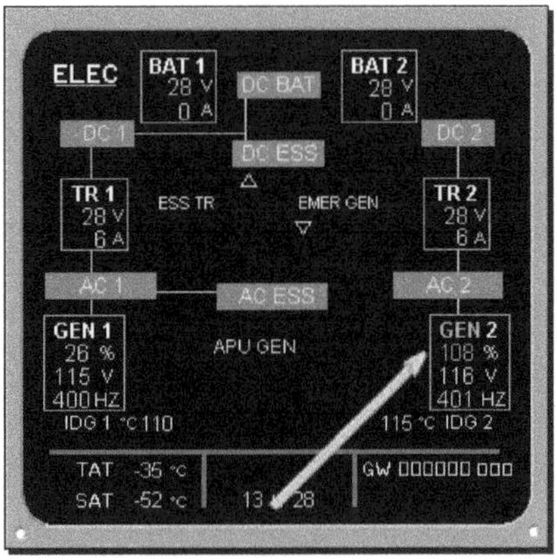

GALLEY FAULT light illuminated on the ELEC panel helping you to locate the switch. (The galley will be switched off to reduce the load.)

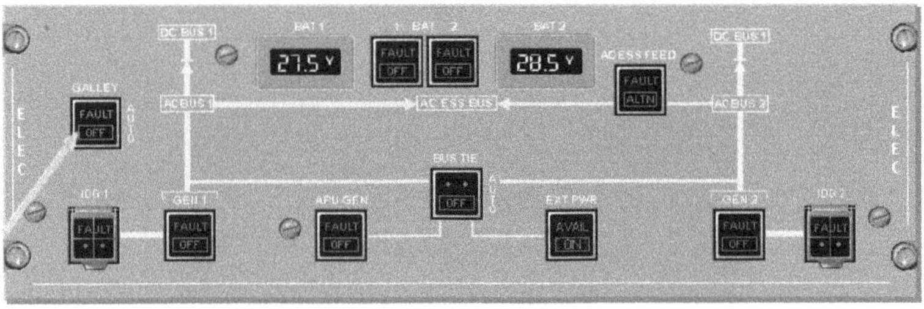

After performing ECAM ACTIONS, GALLEY is OFF and the system trends to normalize.

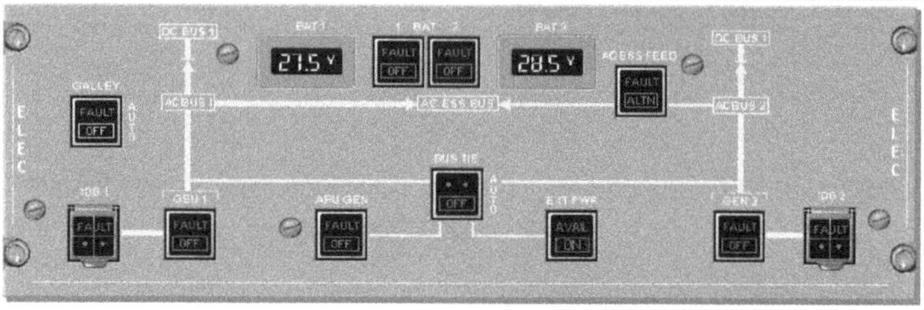

Notice that there is no FAULT light in the GEN 2 pb sw because the generator is still powering the AC bus 2. Now, you can see, on the ECAM ELEC page:

- GALLEY SHED indication has appeared

- GEN 2 load is back to a normal value.

In this next step we will look at an AC BUS 1 fault. Assume we are in flight, with no previous faults. Autopilot 1 and autothrust are engaged. Let's first review the ECAM ELEC page. As you already know, AC bus 1 is normally supplying:

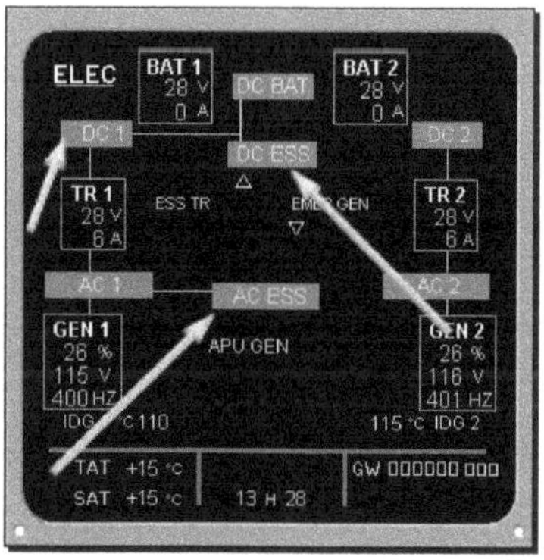

- AC ESS bus and
- DC bus 1 which in turn feeds DC ESS bus.

If you loose AC bus 1, these same buses will be lost. The essential buses can be recovered using the AC ESS FEED pb sw. We will demonstrate this point. Let's activate the failure!

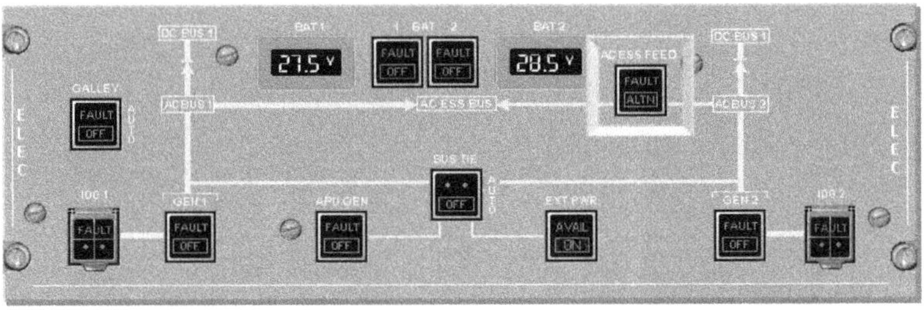

Suddenly the MW alarm is triggered...

As you notice, the first consequence is that the Captain's PFD, ND and E/WD are blank. The SOP recommend, in such a case, that the First Officer should fly the aircraft while the Captain does the ECAM action.

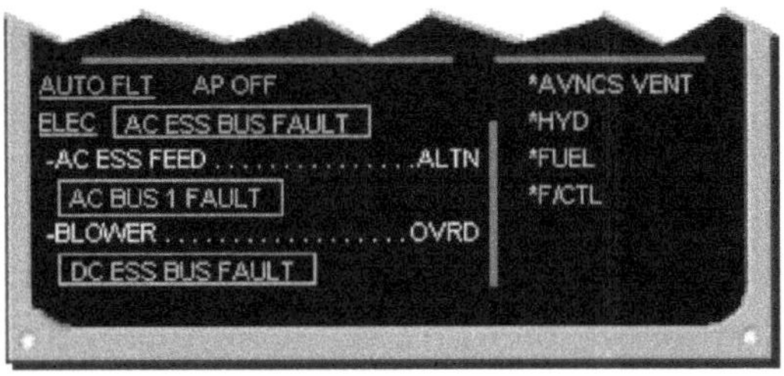

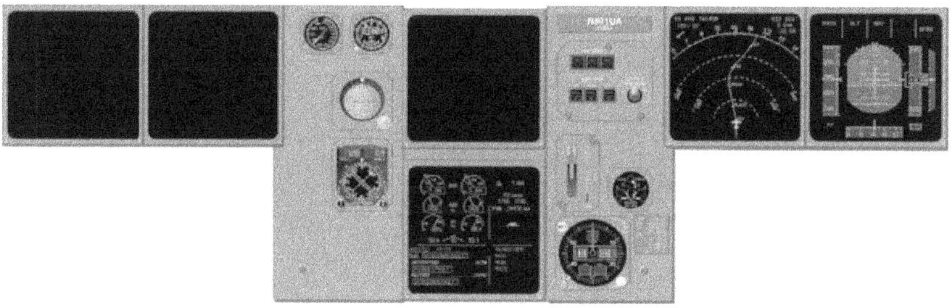

The E/WD is now displayed on the lower ECAM. You can read that the autopilot has disengaged. Let's deal with this, first. To continue the procedure, you have to clear this line. The line concerning the autopilot has disappeared. We can also see that the AC bus 1 fault has led to AC and DC ESS bus failures.

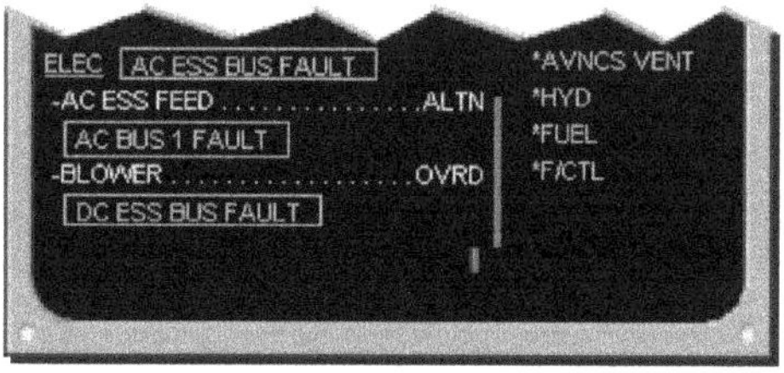

The first line in the procedure shows us how to recover the essential buses. You will have to switch the "feed" of AC ESS bus from AC bus 1 to AC bus 2. The PF will ask you to perform the ECAM actions. Notice on the ELEC panel, the FAULT light on the AC ESS FEED pb sw.

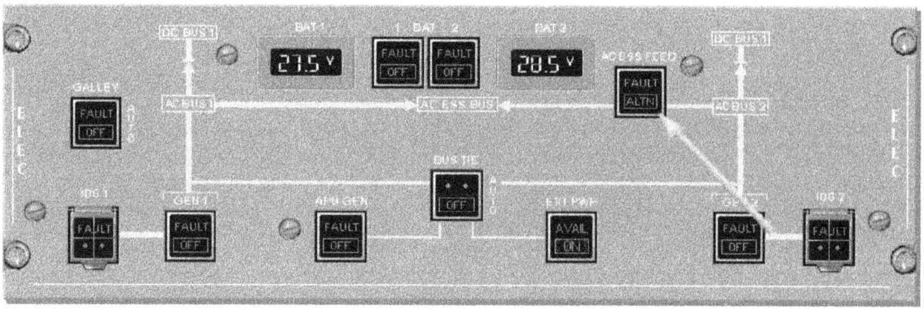

Let's see what happen after ECAM ACTIONS!

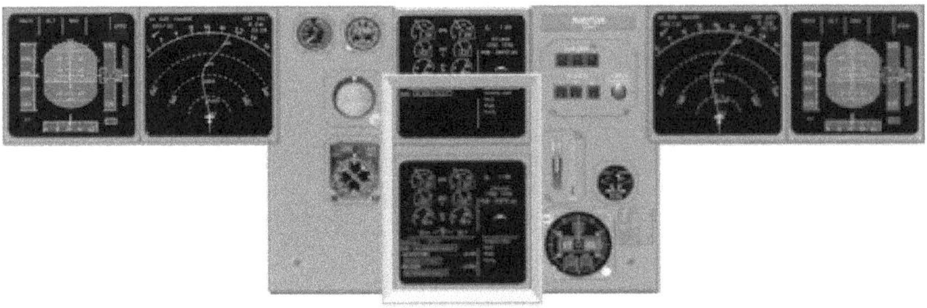

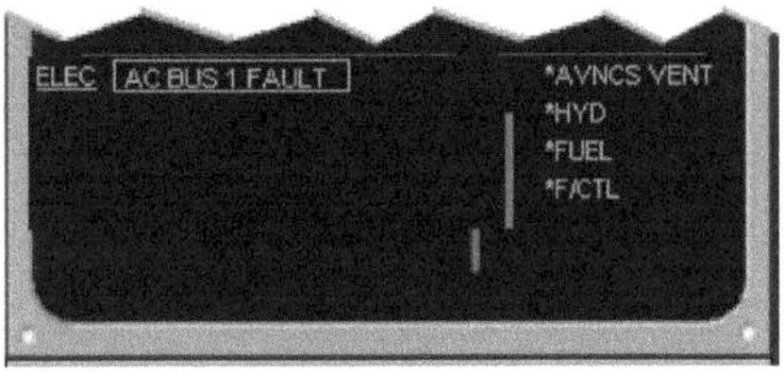

The Captain's PFD and ND are recovered after a few seconds. The E/WD is back on the upper ECAM. The ECAM ELEC page is automatically displayed on the lower ECAM. You can see on the ECAM ELEC page that both essential buses are now recovered. For that, AC bus 2 is now supplying the AC ESS bus which in turn, via the ESS TR, supplies the DC ESS bus.

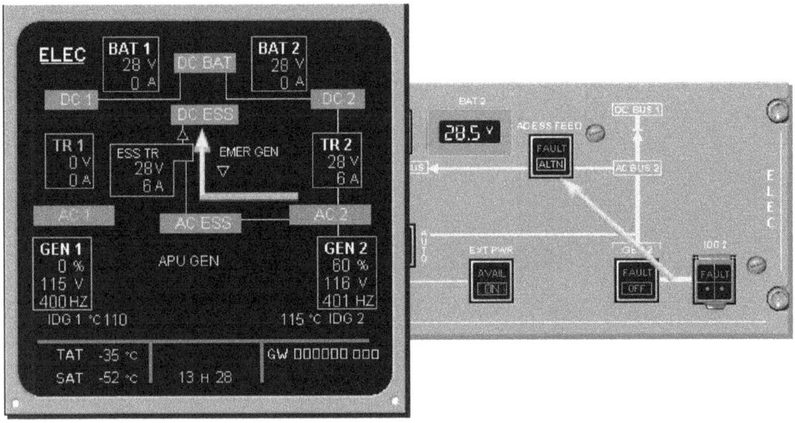

On the ELEC panel, the FAULT light on the AC ESS FEED pb sw has been replaced by a white ALTN light. This means that the AC ESS bus is supplied by its alternate source, AC bus 2. Notice that DC BAT and DC 1 buses are automatically supplied by DC bus 2 after a short delay. The other steps of the procedure have disappeared with the recovery of the essential buses. The only bus which remains unpowered is AC bus 1. You can see, on the right side of the E/WD, the systems which are affected by the loss of AC bus 1.

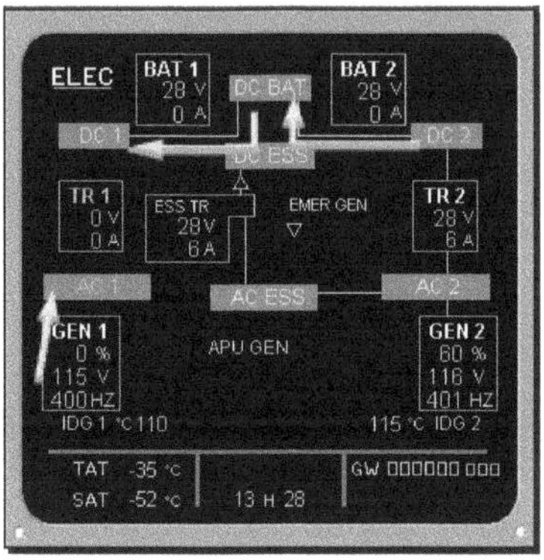

Normally, at this point, you will start to review the ECAM pages relative to the secondary failures. Here, ECAM CAB PRESS, HYD, FUEL and F/CTL pages. As you will see complete procedures during your simulator sessions, let's stop here and go to the next abnormal operation theme.

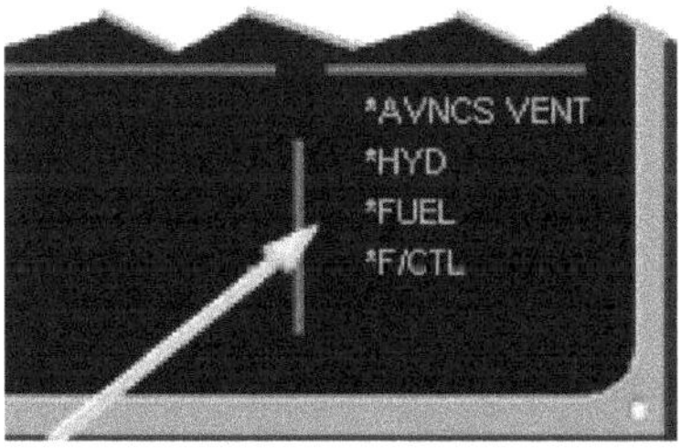

We will now study the ELECTRICAL EMERGENCY CONFIGURATION. As there are a lot of events which happen at the same time in the aircraft, we will present them in a

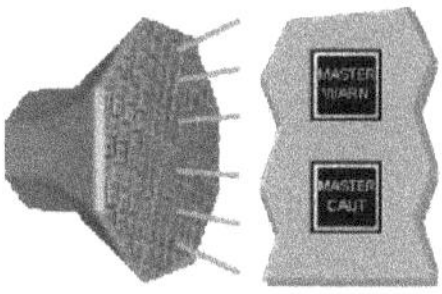

sequential order. Let's assume we are in flight and no previous faults have occurred. Suddenly the MW alarm is triggered…

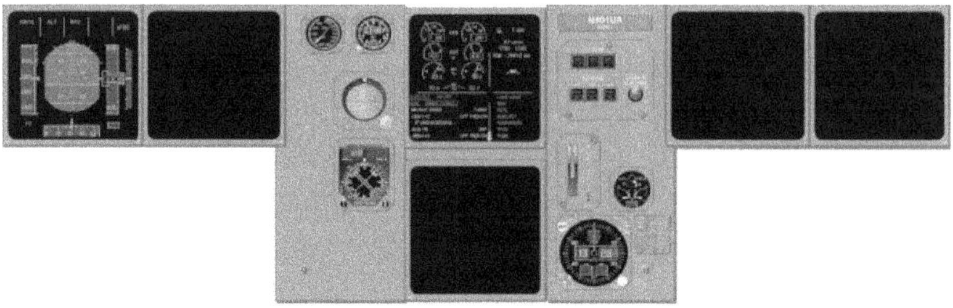

As you can see, the first consequence is that the power supply for the First officer's PFD and ND, the Captain's ND and the lower ECAM is lost. The SOP recommend, in such a case, that the Captain will fly the aircraft while the First Officer does the ECAM action.

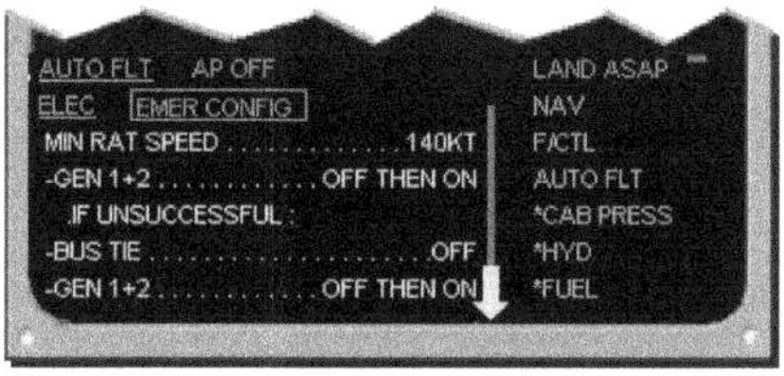

Observe that on the EMER ELEC PWR panel, the red RAT & EMER GEN FAULT light is on meaning that the emergency generator is not yet on line.

Observe that the RAT & EMER GEN light has disappeared. The EMER GEN is now running and supplying the

system. Notice that with the EMER GEN on line, the power supply to the Captain's ND has been restored.

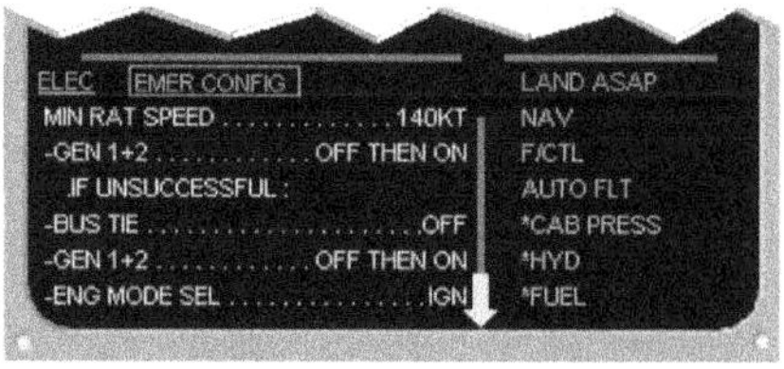

On the E/WD, you can read that the autopilot has disengaged. Let's deal with this, first. To continue the procedure, you have to clear this line. Read the title of the failure. The red EMER CONFIG title means that you are in an ELECTRICAL EMERGENCY CONFIGURATION.

The red message LAND ASAP recommends to land as soon as possible. Observe that, you have to maintain a minimum speed when the RAT is extended. The PF will ask you to perform the ECAM actions. You will try to reset the generators. Here, this will be done for you.

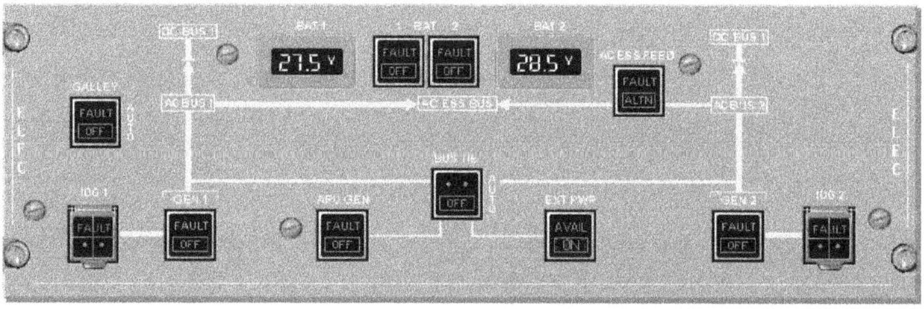

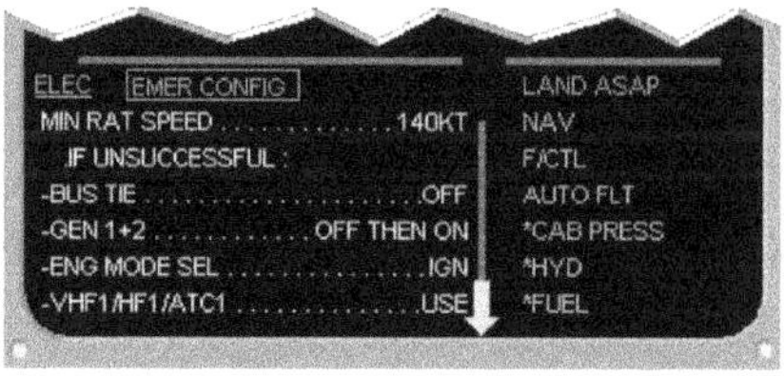

The reset of the generators has not been successful. Before continuing with the procedure, let's look at how to access an ECAM page when the lower ECAM is not available.

The BUS TIE pb sw in the OFF position segregates the two sides of the system. By doing this, a short circuit can be isolated and might give you the possibility to reset one generator. The procedure asks you to try to reset the generators again. Here, this will be done for you.

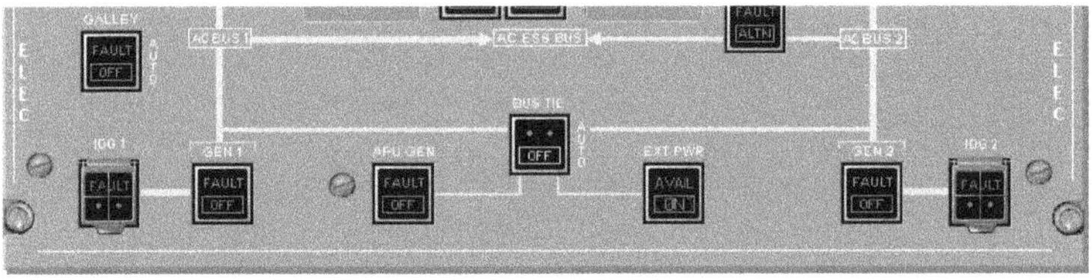

Continue ECAM ACTIONS…

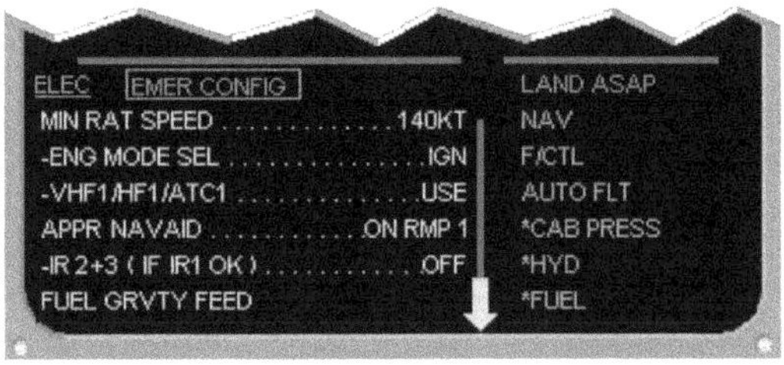

Again, the reset of the generators has not been successful. As the next steps concern other systems, we will stop here and study the indications on the right side of the E/WD. Notice the amber NAV, F/CTL and AUTO FLT system titles. This means that these systems are affected by the failure. Their related procedures are stacked after the current one and will appear with the completion of the EMER CONFIG procedure.

Notice also the visible secondary failures concerning the CAB PRESS, HYD and FUEL systems. There may be other secondary failures that are not yet displayed since there is an overflow indication(green arrow). We will now jump to the end of the EMER ELEC procedure and see how you can view the secondary failures.

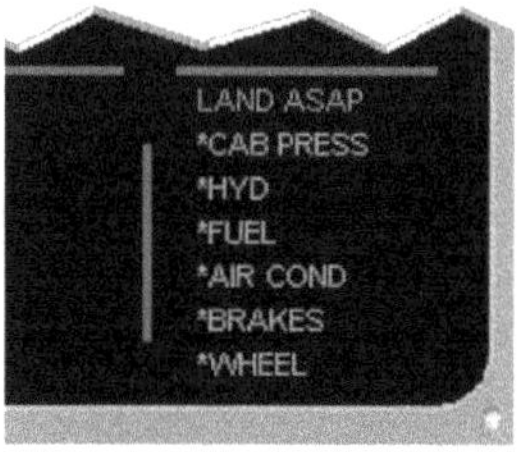

The system titles have disappeared when their procedures are completed. The complete list of the secondary failures is now visible. As the lower ECAM is not available, you will have to manually call the ECAM pages associated with the secondary failures on the E/WD. We will now assume that all the ECAM pages of the secondary failures have been reviewed. As you cleared the last one, memo messages appeared on the E/WD.

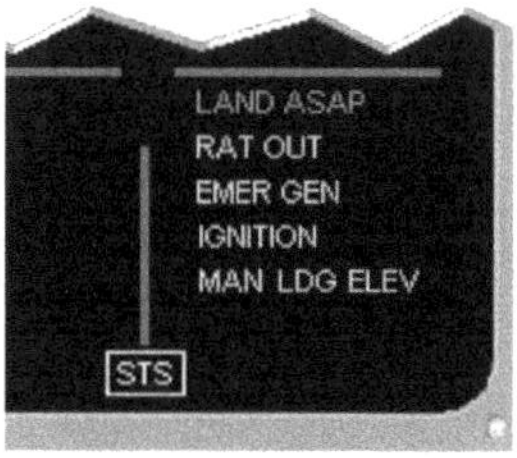

The green RAT OUT memo message confirms that the RAT is no longer stowed. The green EMER GEN memo message indicates that the emergency generator is running and supplying the system. You will have the opportunity to practice the EMER ELEC CONFIG procedure during your simulator sessions. We will now study the case of a RAT stall.

If the RAT stalls, for example at low speed during the approach, the emergency generator is no longer powered and the supply is automatically transferred to the batteries. Observe that the EMER GEN indications have disappeared.

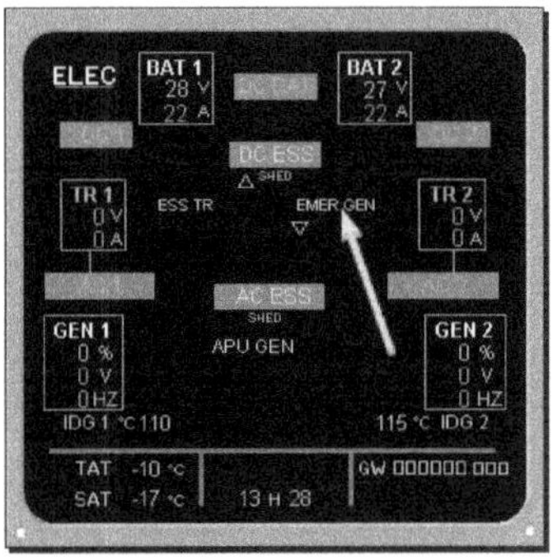

As the battery discharge current is high, the current indications are displayed in amber. The batteries feed the DC ESS bus and, via the static inverter, supply the AC ESS bus.

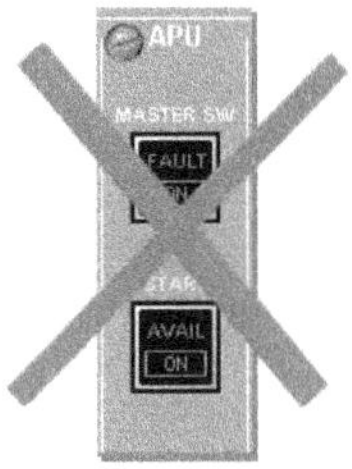

Both SHED ESS buses are automatically shed which is indicated by amber SHED indications below their respective buses. Notice that, during this phase, the APU start is inhibited.

On the ground, the DC BAT bus is automatically connected to the batteries. At low speed, the AC ESS bus is shed leading to the loss of all screens.

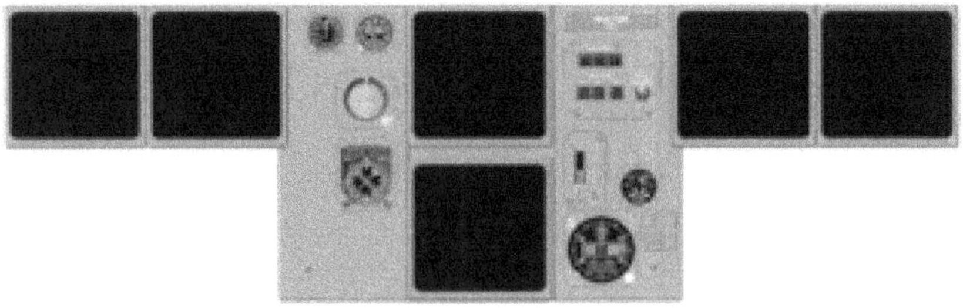

As a result of the electrical failure, the RAT is normally automatically deployed. During the RAT extension, the electrical system is powered from the batteries. The approximate flight time on batteries is 30 min. Let's see what happens if the RAT fails to deploy automatically. If the RAT fails to extend, a specific line appears in the EMER ELEC procedure telling you to manually extend the RAT on the EMER ELEC PWR panel.

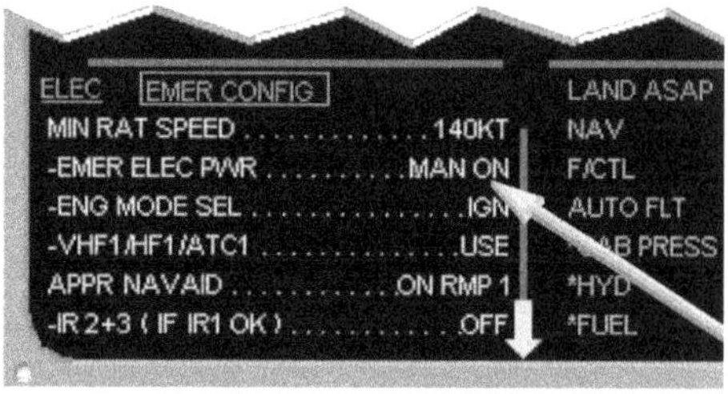

The guard is up. You now have to extend the RAT.

After a RAT extension, if the emergency generator is not working, the red FAULT light stays illuminated on the RAT & EMER GEN indicator. This failure causes the aircraft to be powered only from the batteries.

In case of smoke detection in the avionics bay, the amber SMOKE light illuminates on the GEN 1 LINE pb sw. A Master caution and an ECAM procedure are also triggered.

When the GEN 1 LINE pb sw is pressed generator 1 is disconnected from the system. The avionics smoke ECAM procedure leads to the shedding of the main bus bars. In this case, the electrical system configuration is the same as in emergency except for the fuel pumps which remain supplied.

Air Conditioning Abnormal Operation

This is a demonstration of a pack one overheat in-flight. You are the PNF. Let's assume, you are the PNF. You are in flight and everything is normal, suddenly the MC is triggered! Let's look at the fault indications. A fault message and associated checklist on the E/WD. The ECAM BLEED page has been automatically called to show amber indications. A FAULT

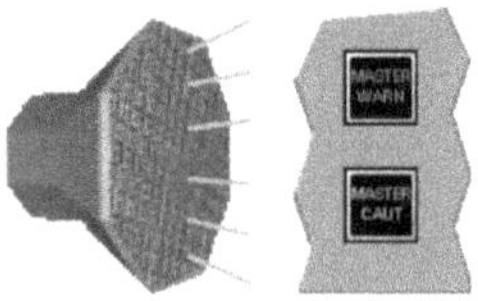

light has illuminated on the Air Conditioning Control Panel.

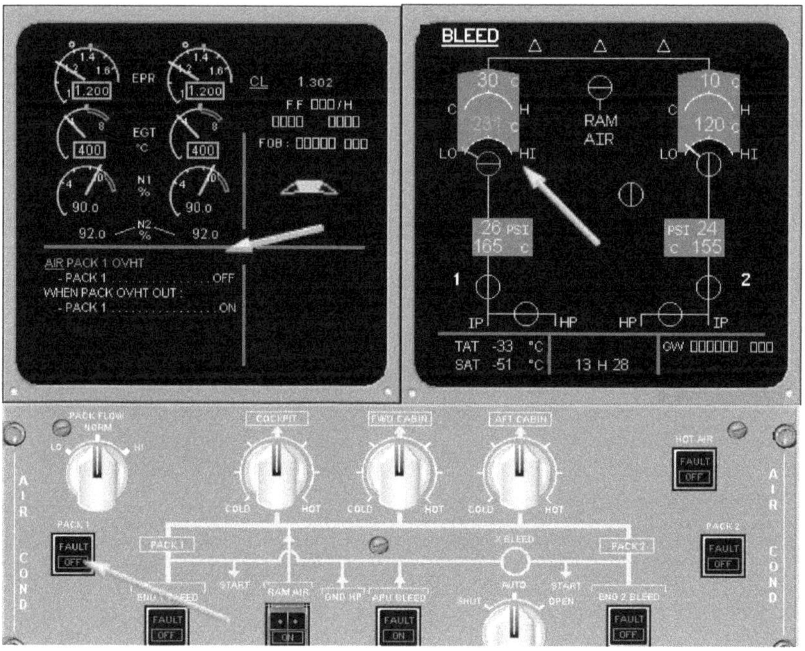

On the E/WD, read the title of the failure. Before you begin, notice that the pack flow control valve has closed. This occurred automatically when the overheat was detected to protect the pack from damage. The valve color indication is amber because the valve position disagrees with the switch position. The compressor outlet temperature is also amber because the temperature limit has been exceeded. Notice also that since the pack valve has closed the supply line from the pack to the mixer unit has turned amber. This happens because there is no air flow from the pack to the mixing unit.

The first step on ECAM directs you to turn off PACK1. This is to match pack switch and pack valve position and prepare to reset the pack. Note that the fault light on the Pack pushbutton switch is illuminated to help you locate it and to indicate the overheat condition.

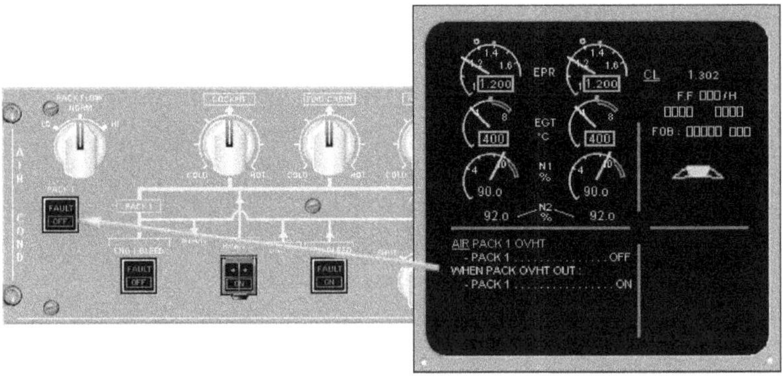

When pack one is turned off, an OFF light illuminates in the switch and the pack valve indication turns green showing valve / switch agreement. Notice that the FAULT light is still on.

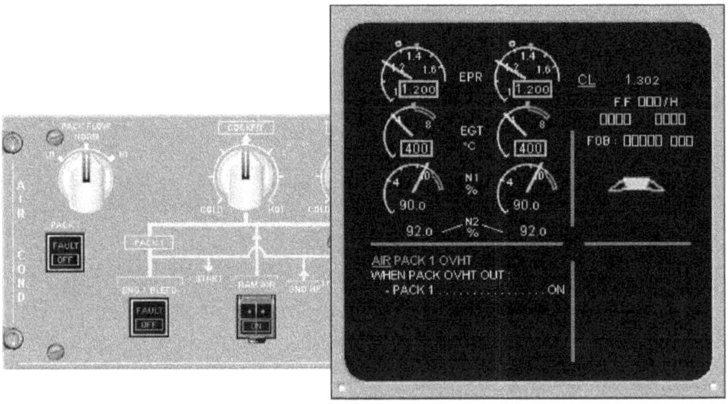

The completed procedural step disappears from the E/WD. Let's move on to the next step of the procedure, which has two parts. The first part, a condition line, is not an action but analysis. We are to determine if the pack overheat is out. Therefore, accomplish the next ECAM action by turning Pack 1 back on.

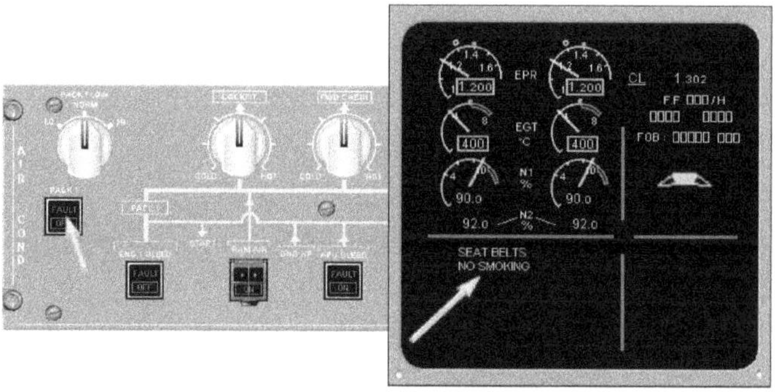

This is an example of an abnormal procedure. We will use a Trim Air System Fault to illustrate these procedures. You are in cruise and all the systems are functioning normally. You are the PNF. When the ECAM system detects a failure, the appropriate alerts are generated, and the system screen called up. Because this is only a crew awareness, neither the Master Caution nor the Master Warning Lights illuminate.

The failure message is read "Conditioning. Trim Air System Fault". The ECAM procedures are now followed. In the case of a

Trim Air System Fault, there is no action required by the crew. The next step in this example, after checking with the other pilot, is to clear ECAM. Because this was just a crew awareness and no action was required by the crew, no STATUS page is displayed. The CRUISE page is now displayed and the CLEAR lights are out. The ECAM procedure for a trim air fault is complete.

Now, let's look at some other Abnormal indications associated with the Air Conditioning system. Should an ECAM or QRH procedure require the use of the RAM AIR switch it is operated by lifting the guarded flap and pushing the switch. Let's Select the RAM AIR switch.

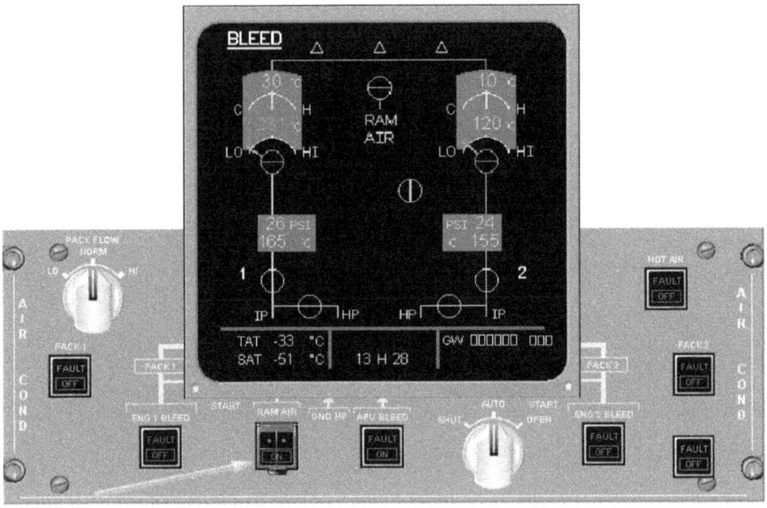

An ON light illuminates in the switch, the valve will open and the indication on the ECAM BLEED page will change to show this. Note that ECAM procedures normally call for the aircraft to be below FL100/Minimum En-route Altitude before operating the RAM Air valve, since the aircraft will depressurize.

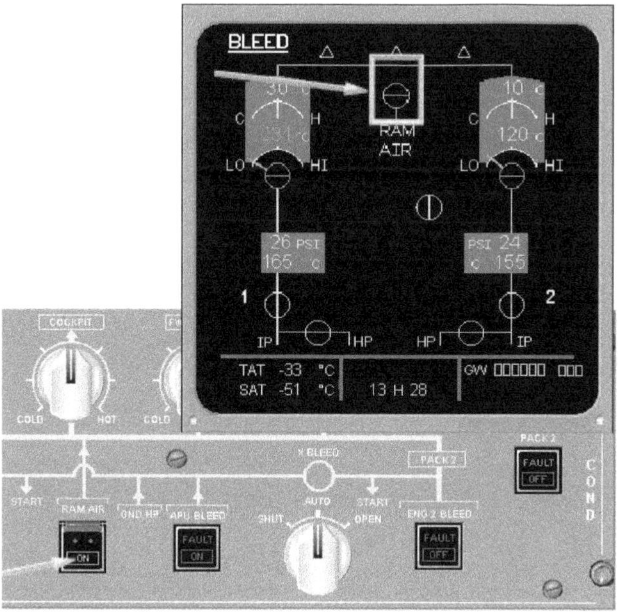

Let's look at another abnormal indication. The HOT AIR FAULT light illuminates amber when a duct overheat is detected. If this occurs the hot air pressure regulating valve and the trim air valves automatically close.

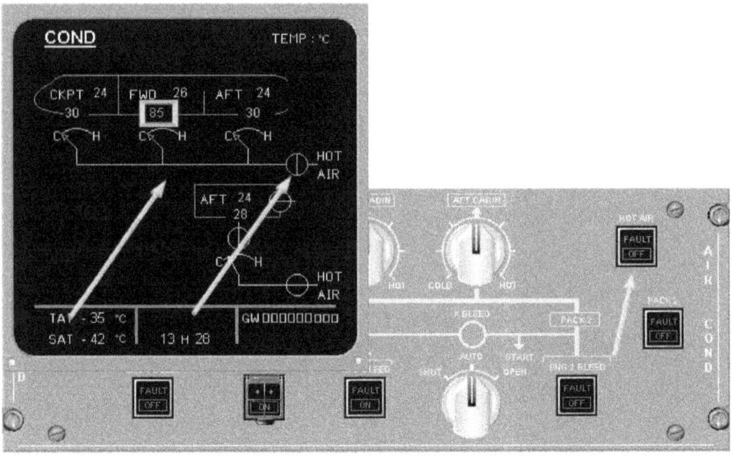

The Zone Control Computer, with two channels, normally operates without any indications to the pilots. However, should a fault occur there are back up modes of operation with associated indications.

In the first case, should a primary channel failure occur the secondary channel takes over, and optimized temperature regulation is lost. An ALTN MODE message appears and the HOT AIR and TRIM AIR valves close.

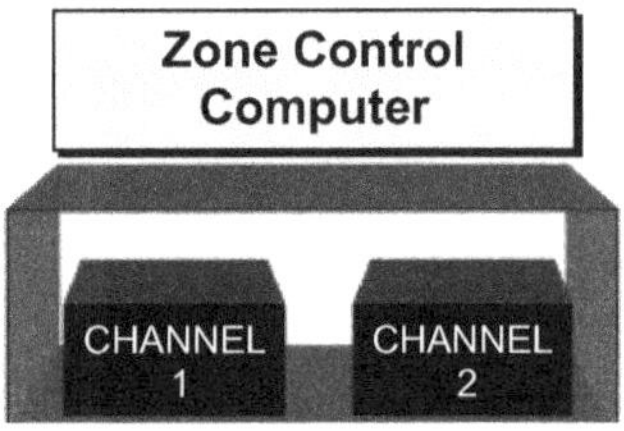

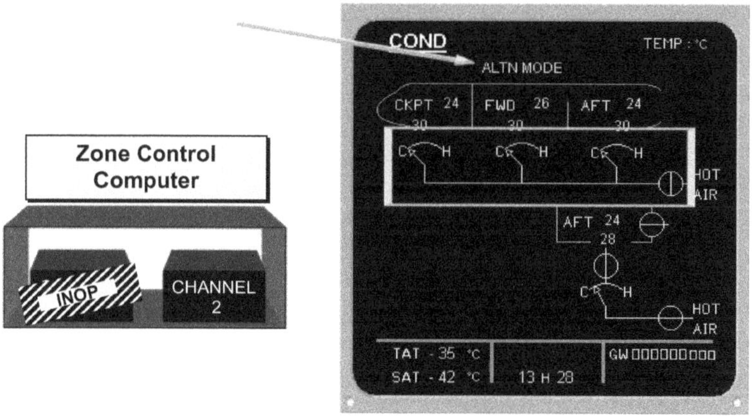

In the second case if a dual channel failure occurs there is no zone temperature regulation. The packs deliver a fixed temperature and the message PACK REG appears on the ECAM COND page. Notice also that there are no indications of zone or duct temperature.

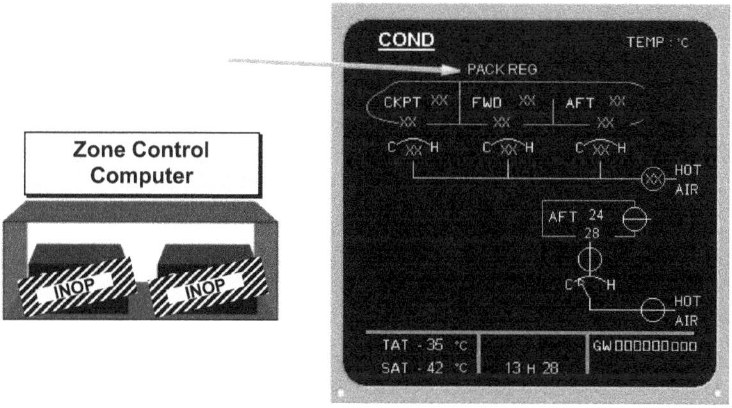

On the STATUS page you are advised that the packs are delivering air at a fixed temperature.

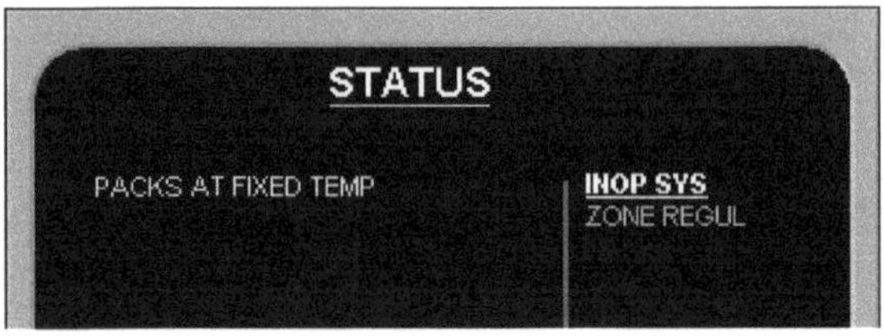

Chapter 2

Abnormal operations II

Aviation Bookstore
aviation step by step

NOTE: *information not valid for real flights. Concepts explained as examples for your study and entertainment. For real flights see the original manuals of the manufacturer AIRBUS.*

Doors Abnormal Operation

In this abnormal module, we will demonstrate a CARGO DOOR fault. Let's assume, you are the PNF. You are at 3000 ft and everything is normal, suddenly the MC is triggered!

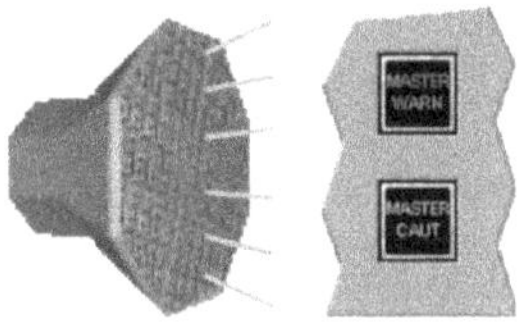

On the E/WD, read the title of the failure.

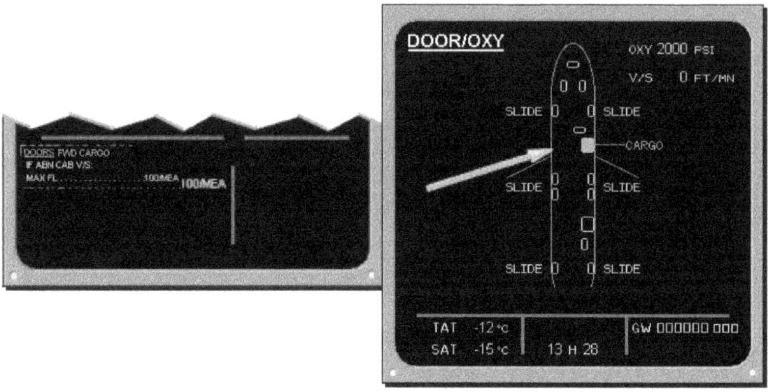

Observe the forward cargo door in amber on the ECAM DOOR/OXY page indicating that the door appears to be unlocked. The procedure requests you to check the cabin V/S. It can be directly read on the ECAM DOOR/OXY page. Here, its value is normal meaning that no depressurization is occurring.

Therefore, no crew action is required and we will proceed with the ECAM procedure by pressing the CLR key. The STATUS page is displayed for review. Notice that there is no approach procedure nor are there any inoperative systems. There is only a limitation in blue MAX FL 100 / MEA.

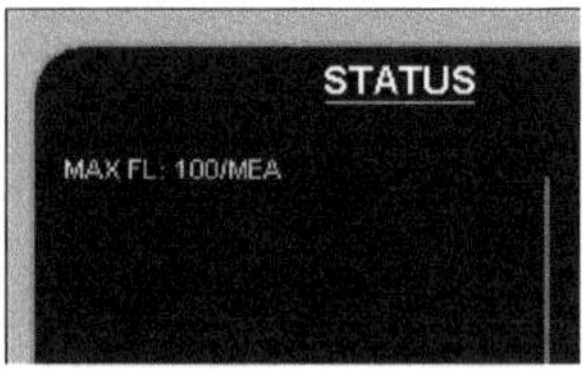

You have seen a CARGO DOOR fault without depressurization. We will now study the same failure but with depressurization. You are in flight and everything is normal, suddenly the MC is triggered!

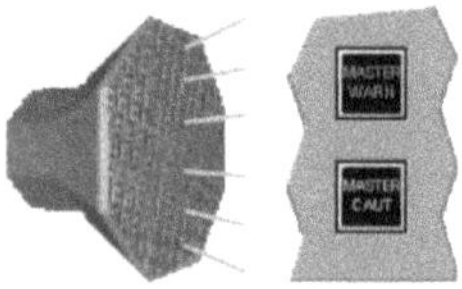

Indications are the same as in the previous procedure except that the cabin V/S is now indicating an excessive value.

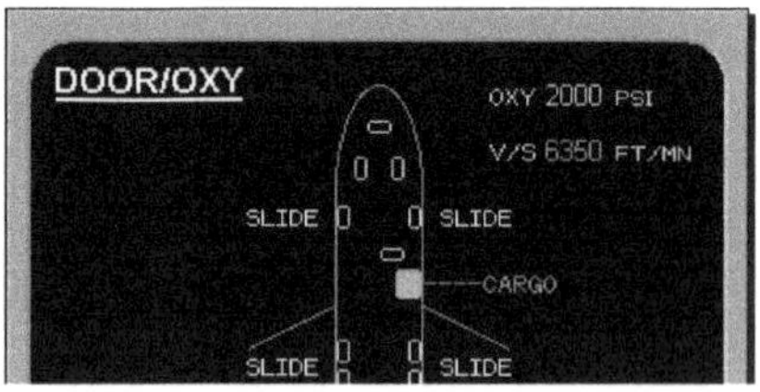

After a few seconds.... The MW alarm is triggered!

Notice that an amber DOOR indication has appeared on the right side of the E/WD due to the overflow, meaning that the DOOR procedure is still present.

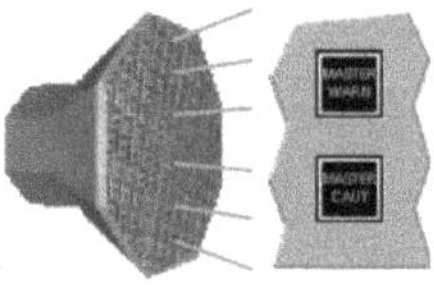

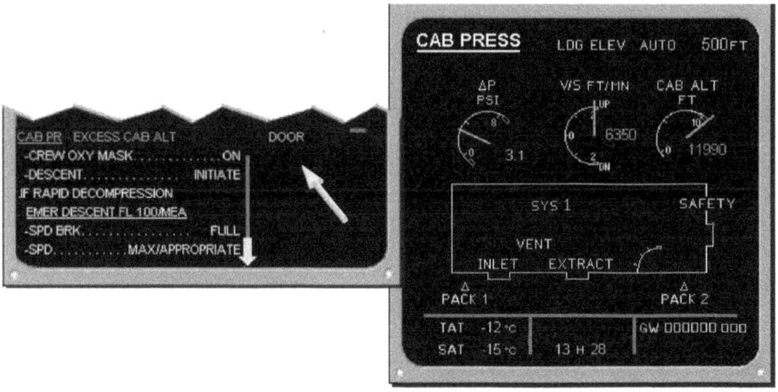

Fuel Abnormal Operation

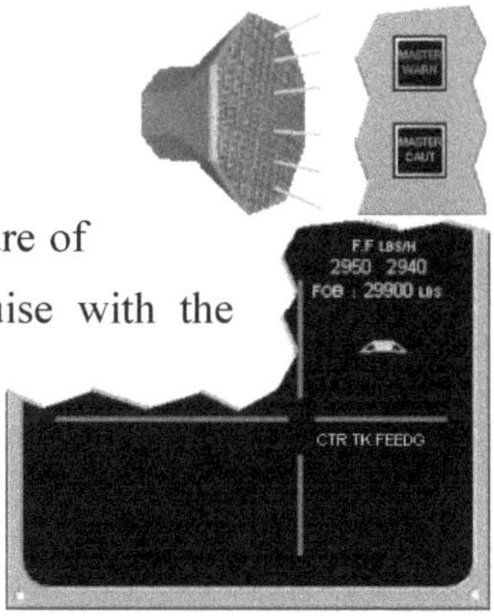

This is a demonstration of a failure of the center tank pumps. You are in cruise with the aircraft operating normally. You are in flight and everything is normal, suddenly the MC is triggered!

On the E/WD, read the title of the failure. The ECAM caution “CTR TK PUMP 1 LO PR” appears amber meaning that center tank pump 1 has failed. Notice that the ECAM FUEL page is automatically displayed.

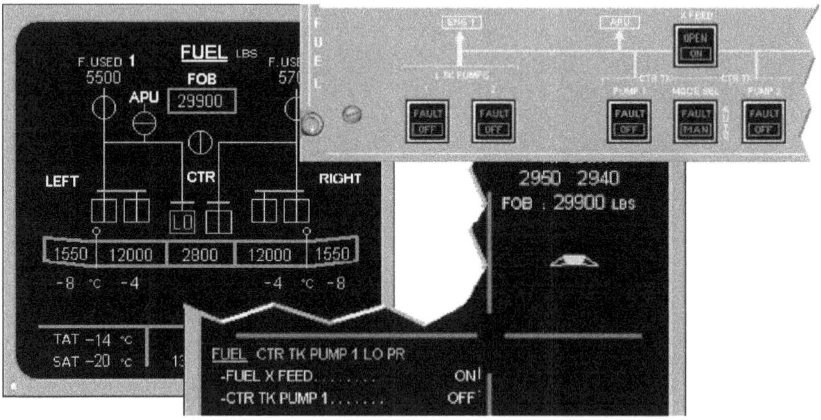

Notice the corresponding indications: On the ECAM page, LO boxed amber means the pump is on, but not producing adequate pressure, and on the FUEL control panel, the FAULT light is illuminated amber on the CTR TK pump 1 pb sw. Let's perform the ECAM actions.

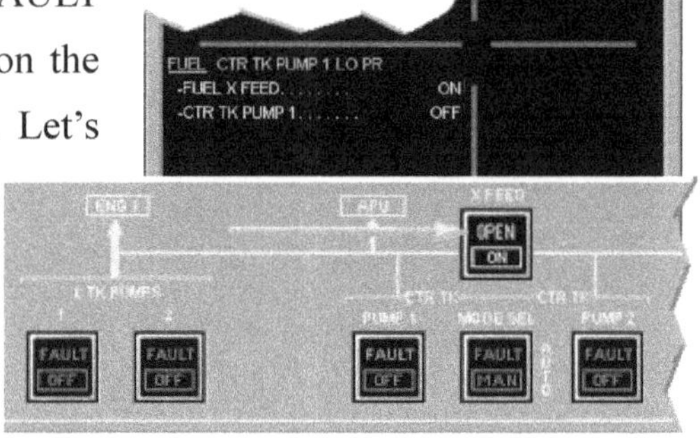

Observe that: the ON light illuminates white on the XFEED pb sw to indicate that the valve starts to open, the crossfeed valve is shown amber during the valve transit The OPEN light illuminates green when the valve is fully open. The crossfeed valve is shown amber during the valve transit.

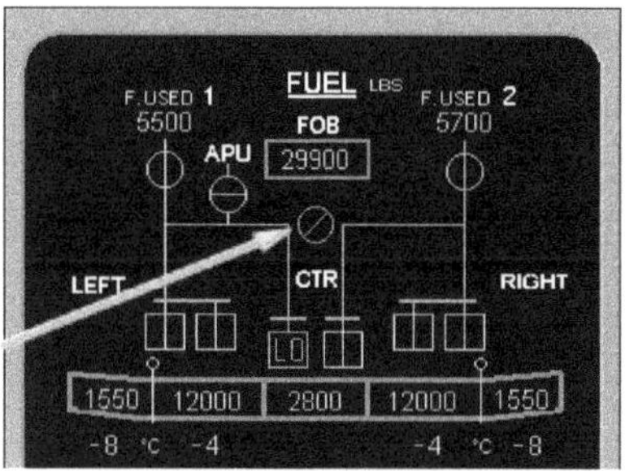

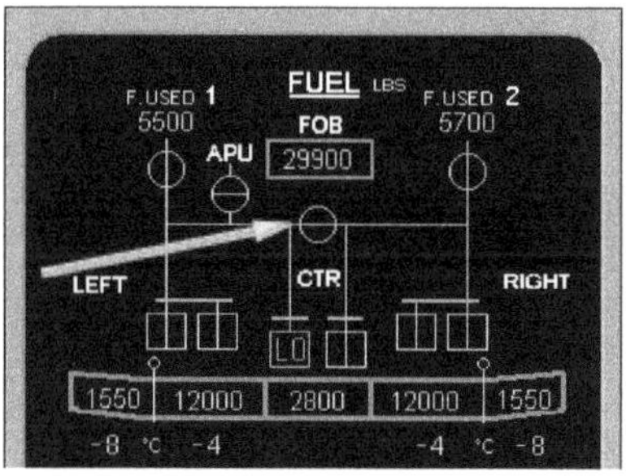

On the ECAM page, the crossfeed valve is in- line green. This allows both engines to be fed from the center tank using center tank pump 2. **Note:** If you do not open the XFEED valve, engine 1 receives fuel from the left inner tank and an imbalance will occur. On the CTR TK pump 1 pb sw, the amber FAULT light disappears, and is replaced by the white OFF light, indicating that the pump has been turned off.

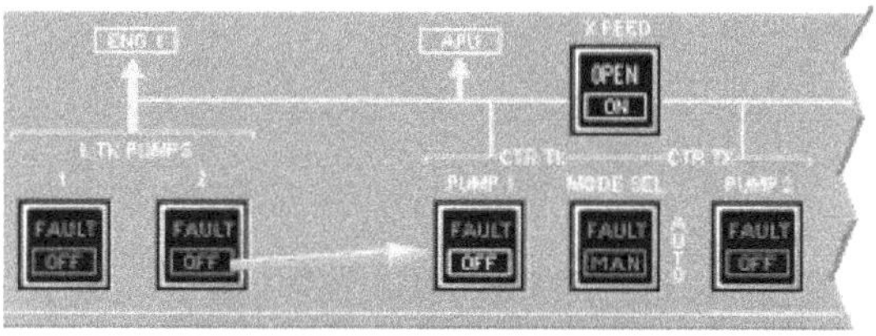

On the ECAM FUEL page, the LO indication disappears and is replaced by an amber cross-line, meaning that center tank pump 1 has been manually switched off.

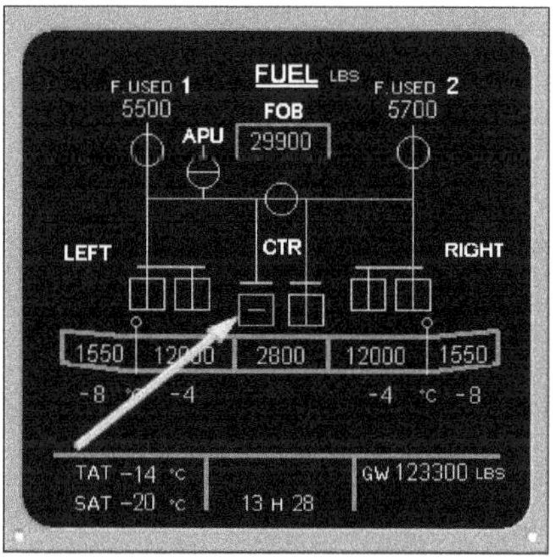

The abnormal procedure is complete, the FUEL system is analyzed. On the STATUS page, center tank pump 1 appears amber in the inoperative system column. ECAM ACTION COMPLETED!

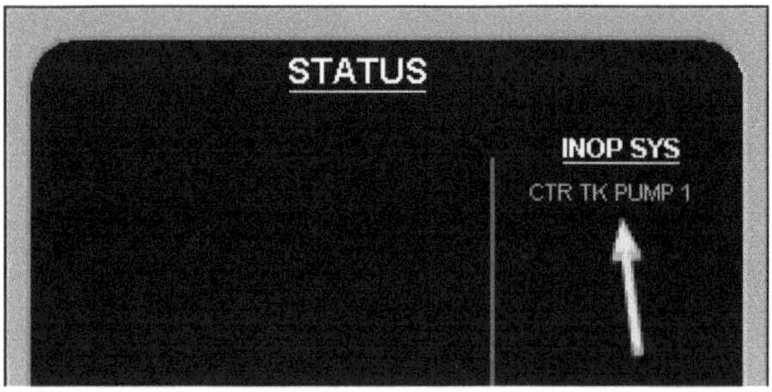

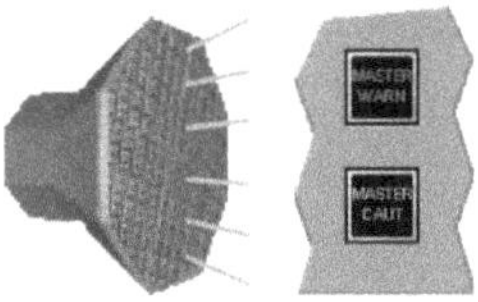

Now you are continuing in cruise. Remember that center tank pump 1 has failed. Let's see what happens if center tank pump 2 also fails. suddenly the MC is triggered.

On the E/WD, the amber message « CTR TK PUMPS LO PR » means that both center tank pumps have failed.

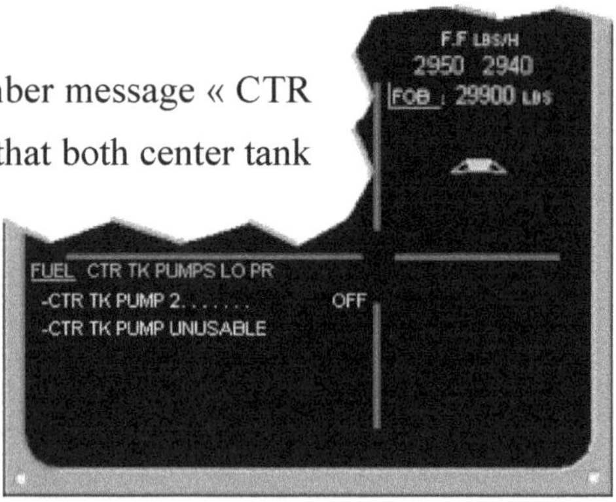

You have the same indications as you had for center tank pump 1 failure: LO boxed amber on the ECAM page, and the amber FAULT light on the FUEL control panel.

Let's perform the ECAM actions.

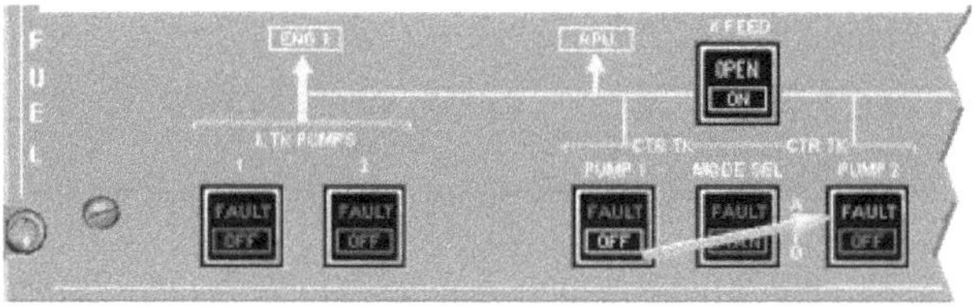

After ECAM ACTIONS (CTR TK PUMP 2 OFF) Observe the following indications :

- On the FUEL panel, both center tank pumps are off,
- On the E/WD, the FOB has a half amber box.

This means that some of the indicated total Fuel On Board is not usable.

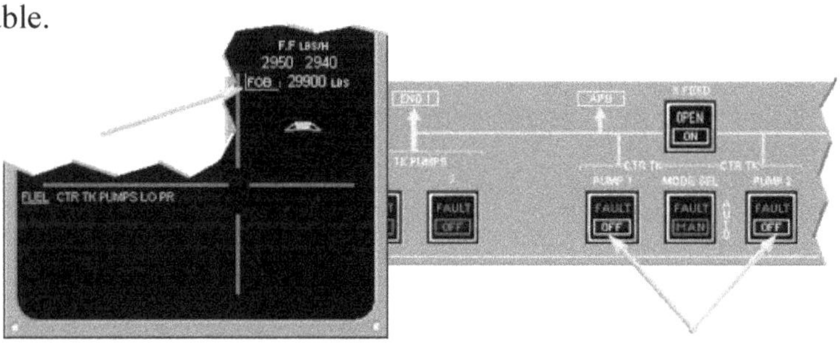

On the ECAM FUEL page:

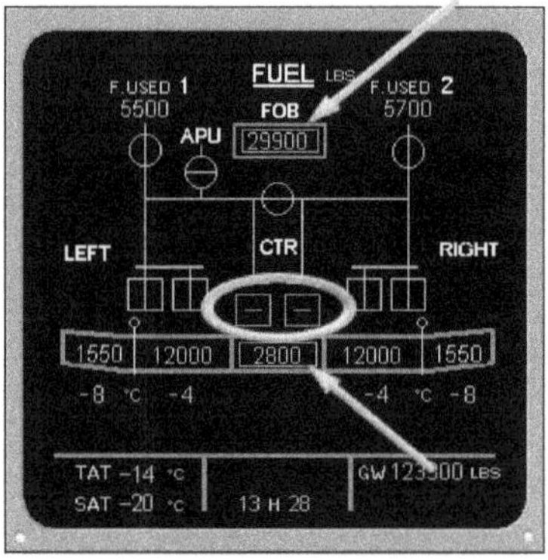

- The Fuel On Board (FOB) value is boxed amber, meaning that not all FOB is usable,

- The center tank pumps are cross-line amber, meaning that they have been manually shut OFF,

- The center tank quantity is boxed amber, meaning that the center tank fuel is not usable.

The STATUS page indicates the following

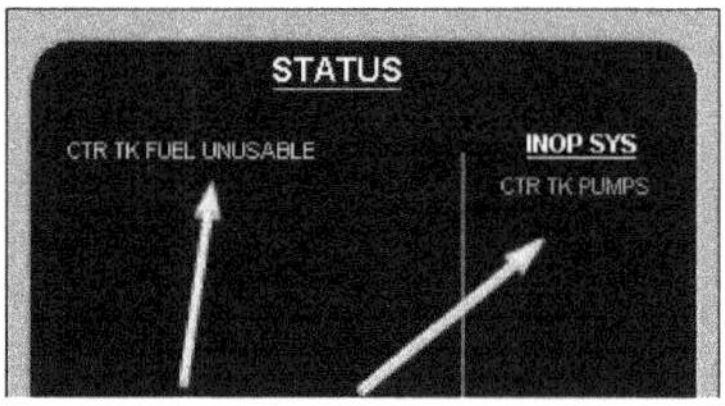

- A green message states that the center tank fuel is unusable

- The center tank pumps are inoperative. This is indicated in amber.

ECAM actions completed.

We learned from the ECAM actions and the indications associated with center tank pump failures that: Losing one center tank pump requires opening the crossfeed valve to prevent a fuel imbalance, Losing both center tank pumps means center tank fuel is unusable. You are still in cruise, all systems are restored and operating normally. We are now going to study the failure of the inner tank pumps. The amber message "L TK PUMP 1 LO PR" on the E/WD means that the left tank pump 1 has failed. The ECAM FUEL page appears automatically with the corresponding pump indication.

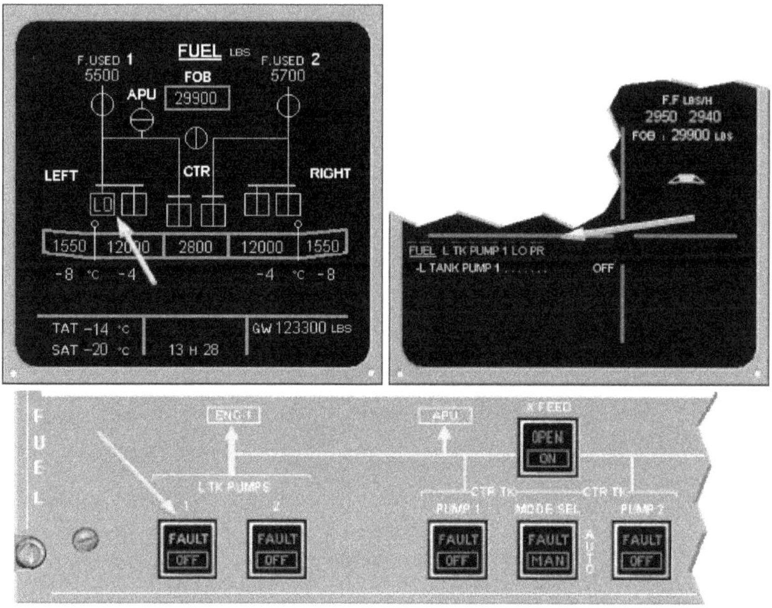

After performing ECAM ACTIONS...

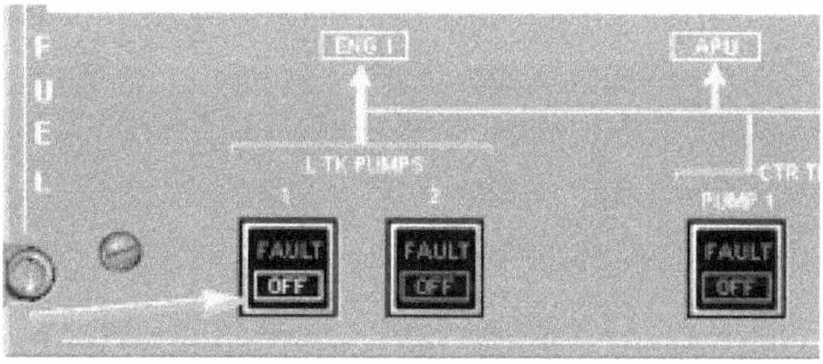

Observe :

* On the FUEL panel, the amber FAULT light in the L TK PUMP 1 pb sw is replaced by the white OFF light.

* On the ECAM page, the amber "LO" indication is replaced by an amber cross-line, indicating that the pump is off.

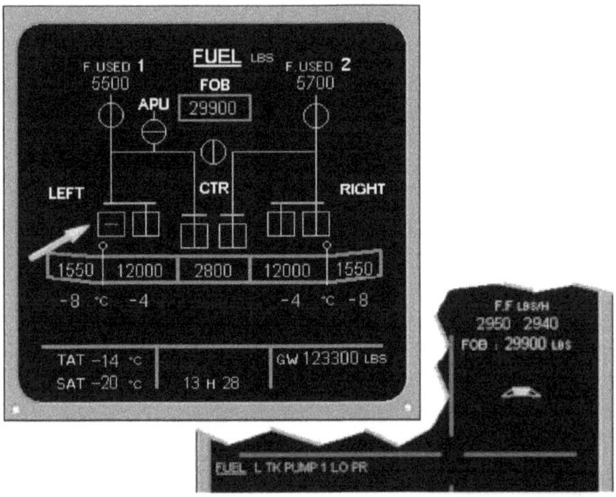

STATUS: The inoperative system is left tank pump 1, indicated in amber.

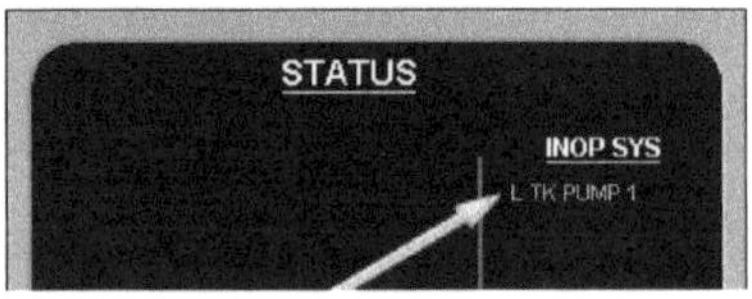

Let's see the next. We are continuing in cruise. Remember that left tank pump 1 has failed. Let's see what happens if left tank pump 2 fails. You are in flight and everything is normal, suddenly the MC is triggered.

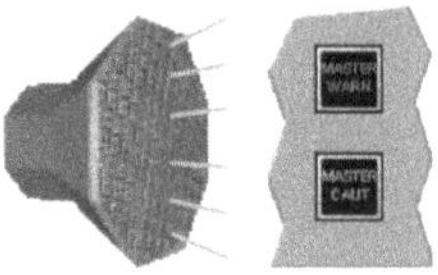

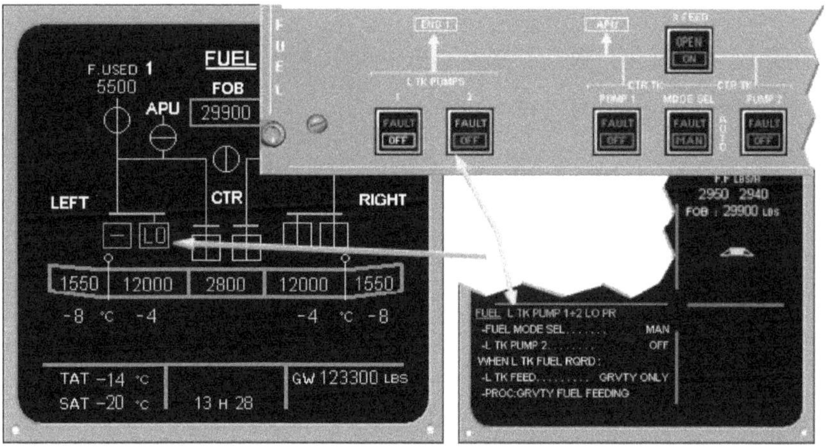

The ECAM procedure requires you to operate the center tank pumps in manual mode. This is to prevent the center tank pumps from being shut off automatically, for example during slats extension.

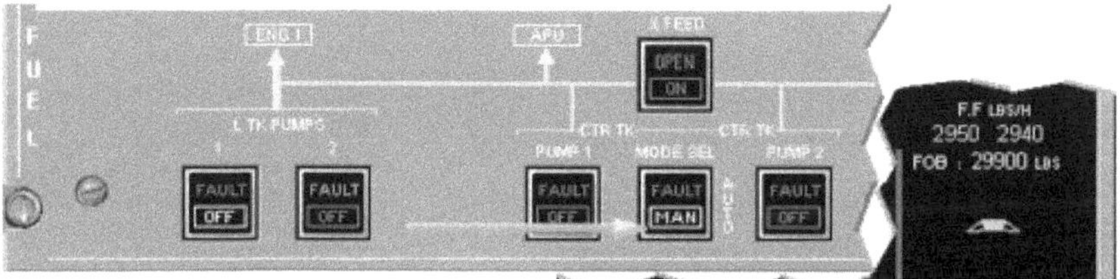

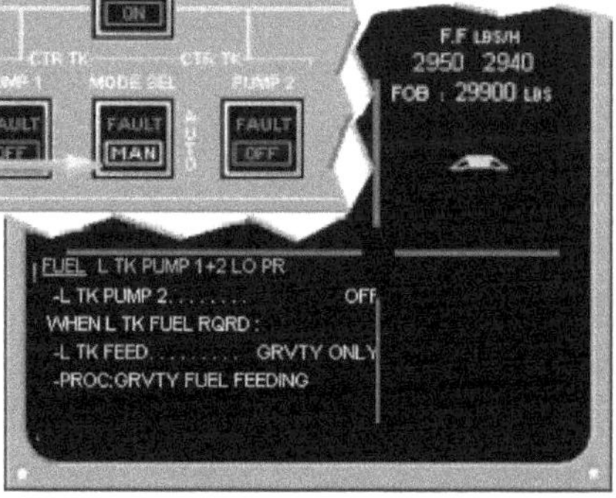

Notice the white MAN light on the MODE SEL pb sw. The center tank pumps are now operated manually. The left tank pump 2 indications are the same as for the left tank pump 1 shown previously.

Let's check the STATUS PAGE. The left tank pumps are inoperative. This is indicated in amber.

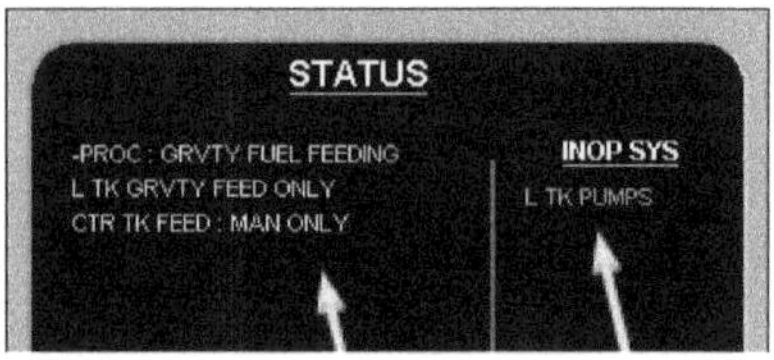

Note: At this time, the center tank pumps are still feeding the engines. When fuel from the left tank is required, refer to your documentation for gravity fuel feeding. ECAM actions completed.

We learned from the ECAM actions and indications associated with the left tank pump failures that:

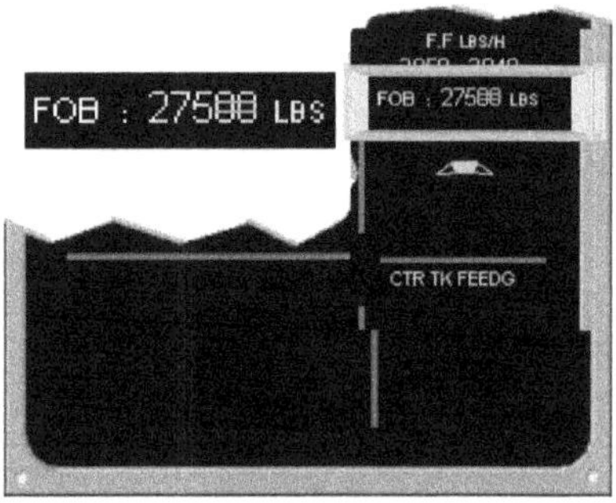

- Losing one inner tank pump is not so critical (no master caution) because of redundancy (2 pumps).

- Losing both inner tank pumps means using gravity fuel feeding if necessary (refer to your documentation).

Now let's see some other abnormal indications that you can have on the ECAM FUEL page. If, on the E/WD, the FOB indication is displayed with two dashes across the two last digits, the Fuel Quantity Indicator

(FQI) is in degraded mode. You can see that the affected tank is the left outer tank. The two amber dashes indicate a loss of accuracy of the indicated fuel quantity.

You can see that the affected tank is the left outer tank. The two amber dashes indicate a loss of accuracy of the indicated fuel quantity.

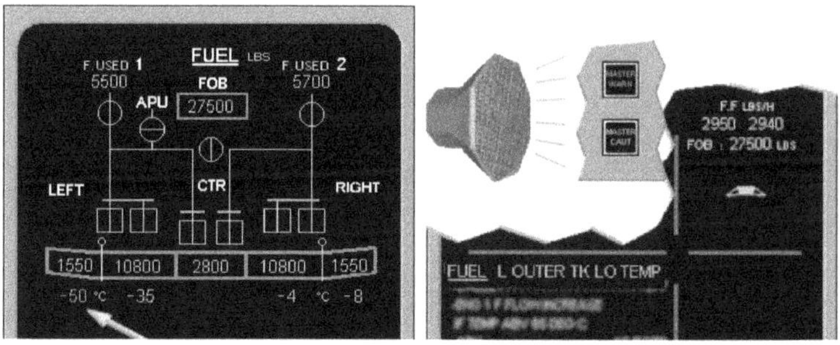

On the ECAM FUEL page, the left transfer valve is displayed in amber, meaning that it is closed when it should have automatically opened. The fuel in the left outer tank is no longer usable.

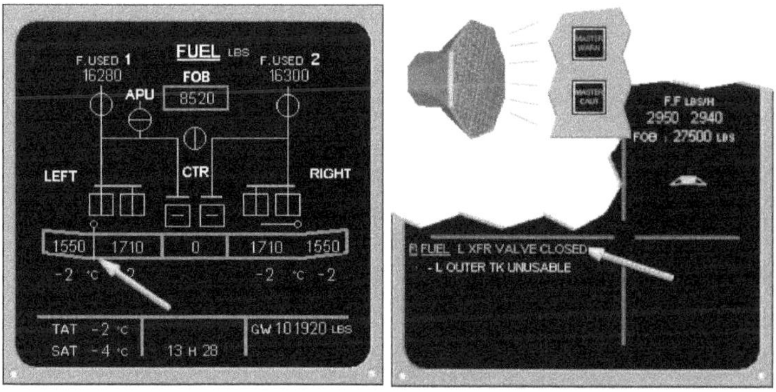

Hydraulic Abnormal Operation

This is a demonstration of an in-flight GREEN Hydraulic abnormal. You are the pilot non-flying. You are in flight and everything is normal, suddenly the MC is triggered!

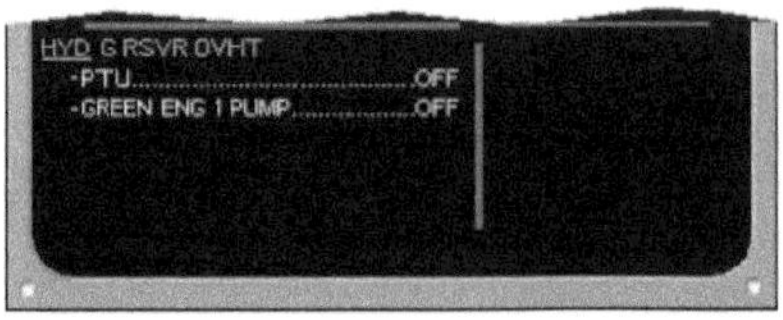

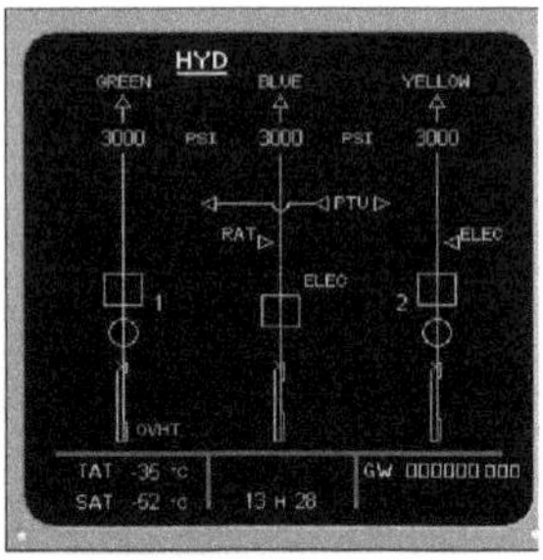

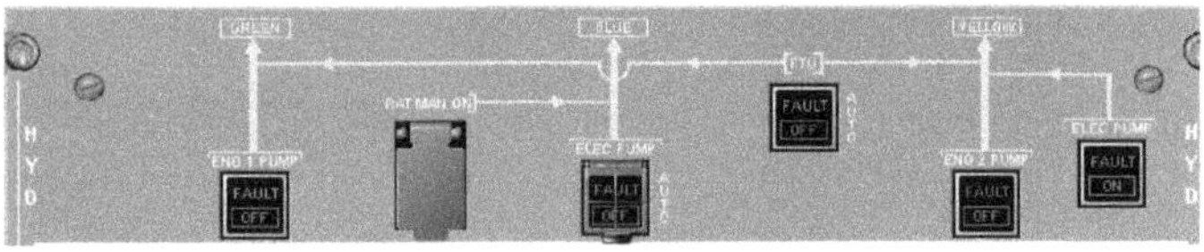

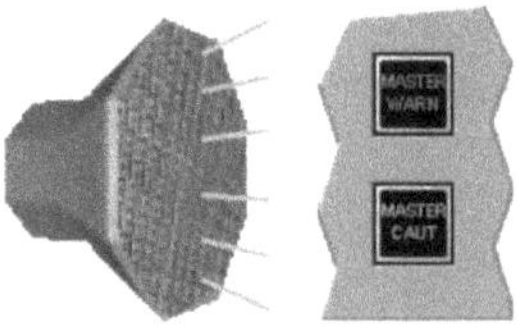

The amber caution and associated procedure are displayed on the E/WD. The HYD ECAM page is automatically presented on the system display. On the E/WD, read the title of the failure. Observe the amber Overheat indication next to the GREEN reservoir on the HYD ECAM page. On the Hydraulic panel, the GREEN ENG 1 PUMP FAULT and PTU FAULT lights are illuminated on the pb sw. These indications provide a cross check of the ECAM and locate the switches to be actuated in the abnormal procedure. Let's perform the first ECAM ACTIONS. Set PUT…..OFF.

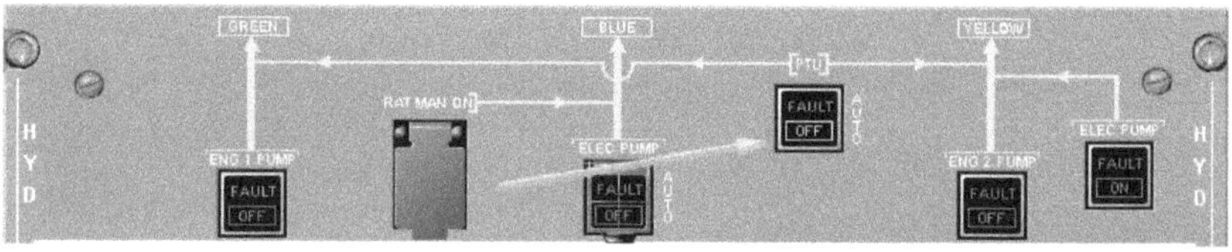

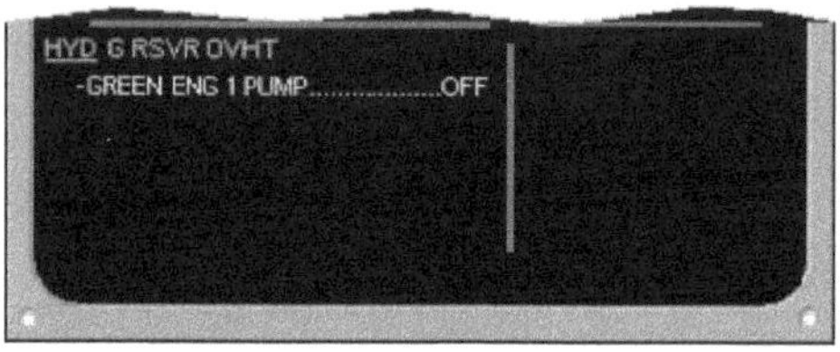

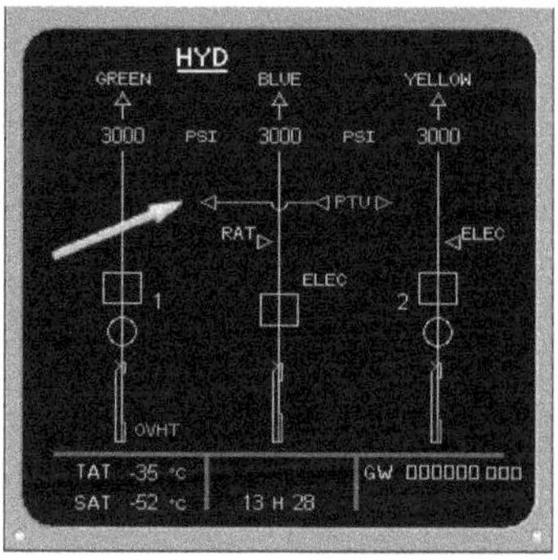

The following indications show that the PTU is switched off: Pb sw white OFF light, ECAM HYD page PTU indication in amber. Let's continue ECAM ACTIONs. Set GREEN ENG 1 PUMP....OFF

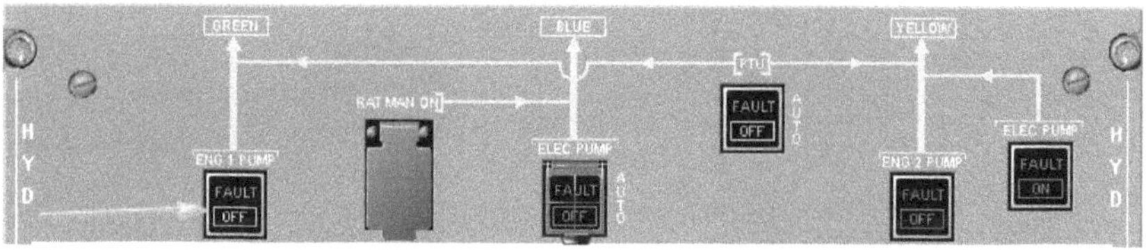

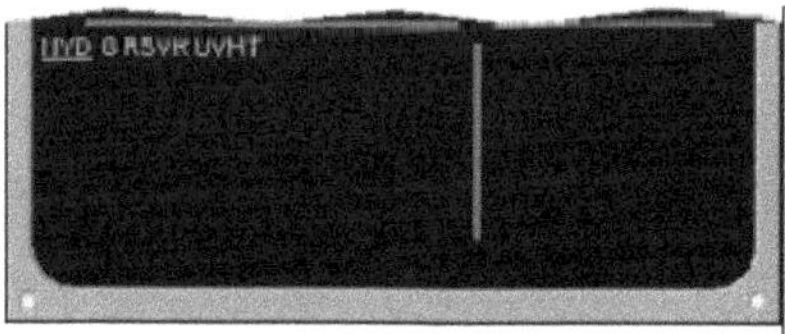

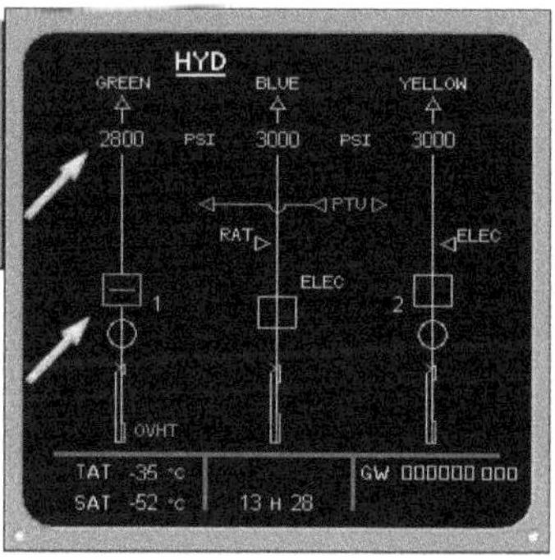

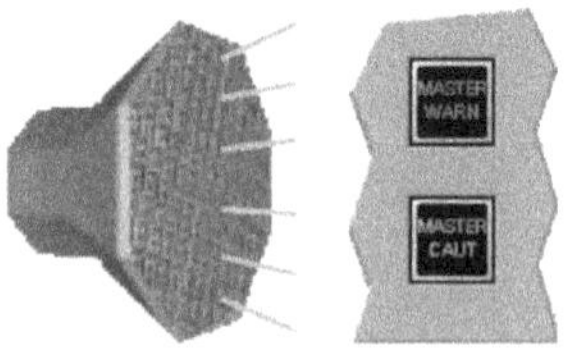

The following indications show that the GREEN ENG 1 pump has been switched off: Pb sw OFF light, ECAM HYD page, GREEN Engine 1 pump has changed from green in-line to amber cross-line. Notice the pressure indication is dropping. Suddenly...

The ECAM Caution, G ENG 1 PUMP LO PR appears amber on the E/WD to indicate that GREEN ENG 1 pump is not producing pressure because you switched it off.

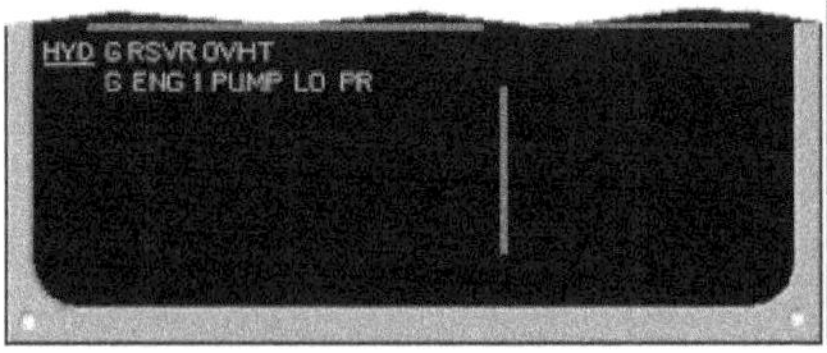

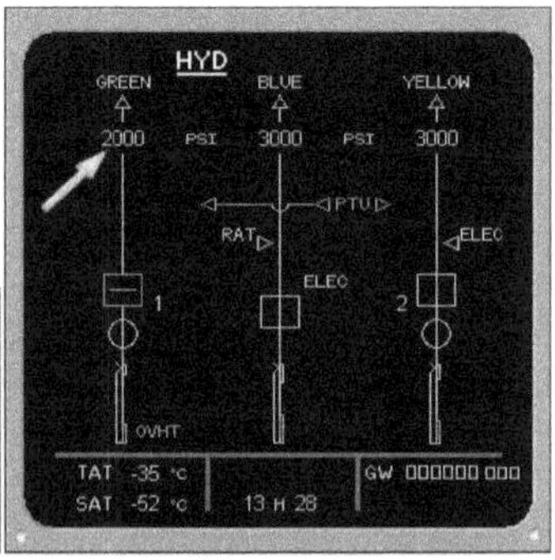

The GREEN System pressure continues to drop and the MC alarm is triggered again...

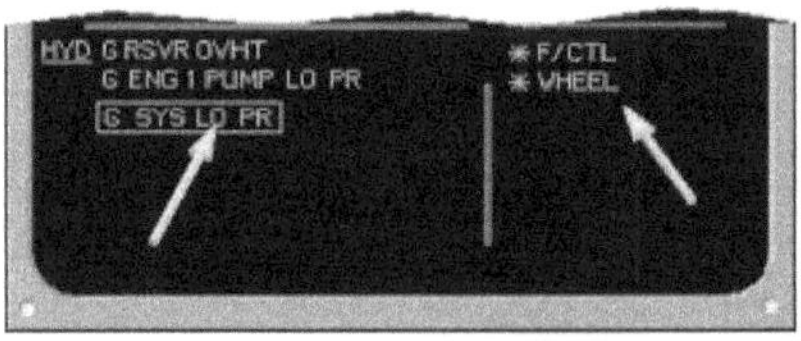

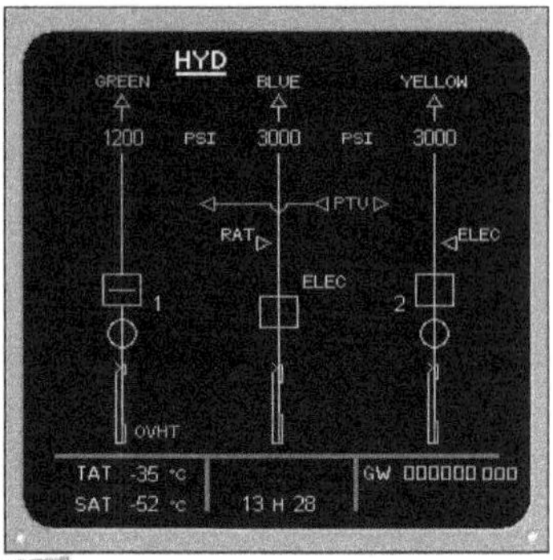

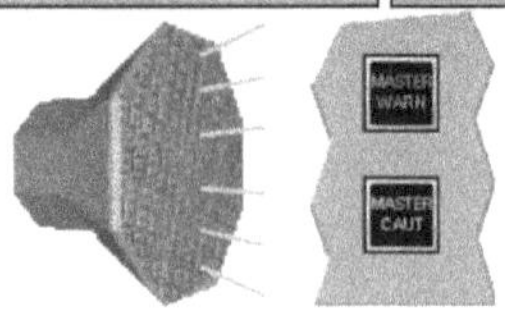

The corresponding ECAM Caution G SYS LO PR is displayed on the E/WD to indicate that the GREEN System is no longer pressurized. The abnormal is boxed to indicate that this is a primary failure with associated secondary failures. At the same time, on the System Display, the GREEN System pressure and line change from green to amber to indicate that the GREEN System is no longer pressurized. In addition, the GREEN system indication changes from white to amber, indicating the loss of GREEN System pressure to the users.

The abnormal procedure is complete. There are secondary failures on F/CTL page and WHEEL page. Let's check it before checking STATUS.

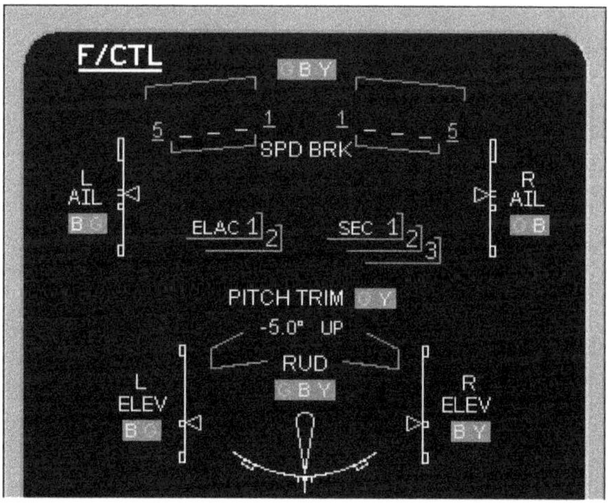

On the ECAM F/CTL page you can see secondary effects of the hydraulic failure. For example, some spoilers are affected by the GREEN System failure.

The ECAM WHEEL page is displayed automatically. Nose Wheel Steering is inoperative due to the failure of the GREEN Hydraulic System.

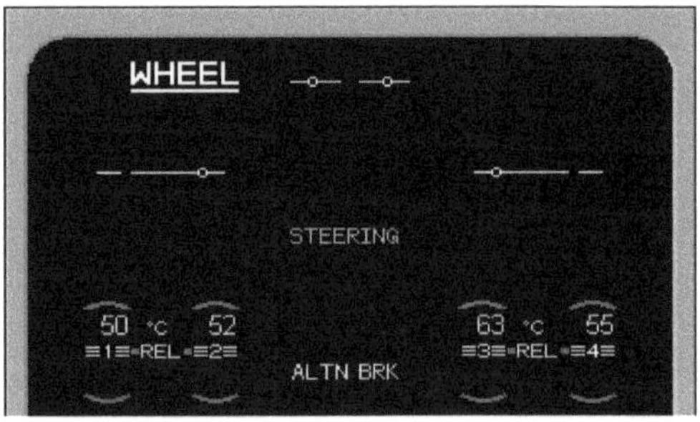

The STATUS page is displayed showing: approach procedures, inoperative systems and aircraft information. ECAM ACTIONS COMPLETED!

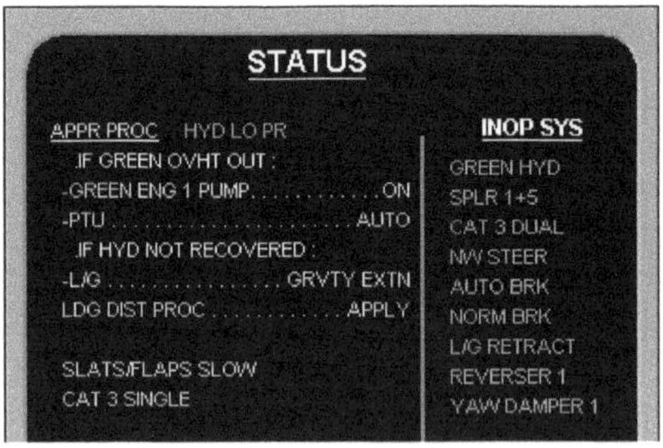

Let's continue flying. The GREEN Hydraulic System is restored. The STATUS page shows NORMAL for a few seconds. Then the ECAM System Display automatically reverts to the CRUISE page. Let's look briefly at another reservoir problem which can result in a similar situation.

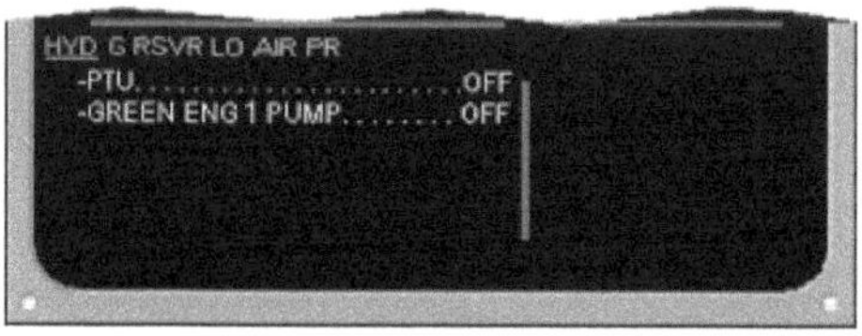

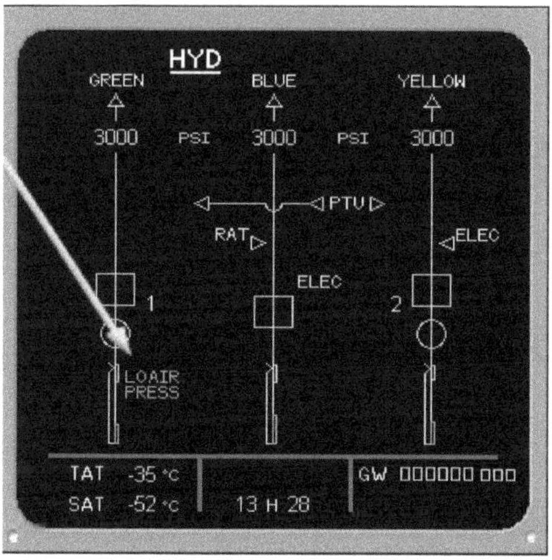

This failure is indicated by:

- HYD G RSVR LO AIR PR on the E/WD,

- and the display of the amber "LOW AIR PRESS" message next to the affected reservoir.

Notice this procedure and the indications are similar to those in the previous example (G RSVR OVHT). With this malfunction, it is also possible to recover the system using the APPR PROC. We will assume the ECAM actions are complete.

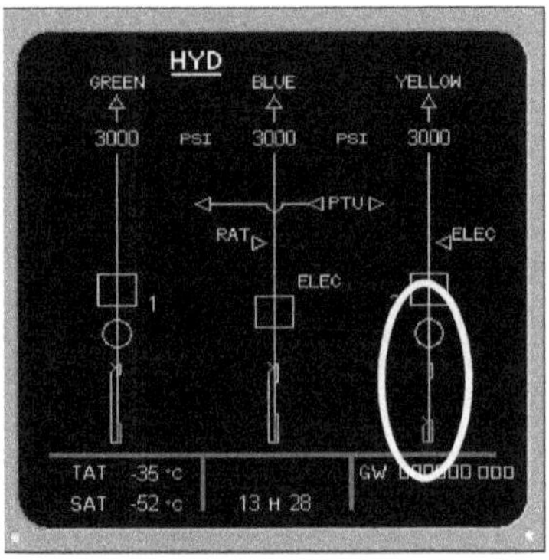

This is a demonstration of an in-flight YELLOW RESERVOIR LOW LEVEL. You are the PNF. You are in cruise, everything is normal. Suddenly the MC is triggered! On the E/WD, read the title of the failure.

Observe the reservoir level indicator has changed to amber because the contents are below the warning level. Let's perform the first ECAM ACTIONS. Set PUT to OFF. Notice the the PTU indication on the ECAM. It's has changed to Amber.

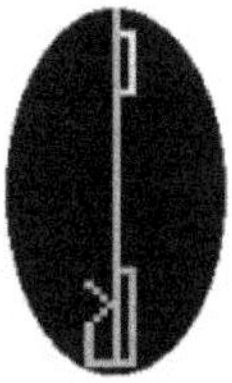

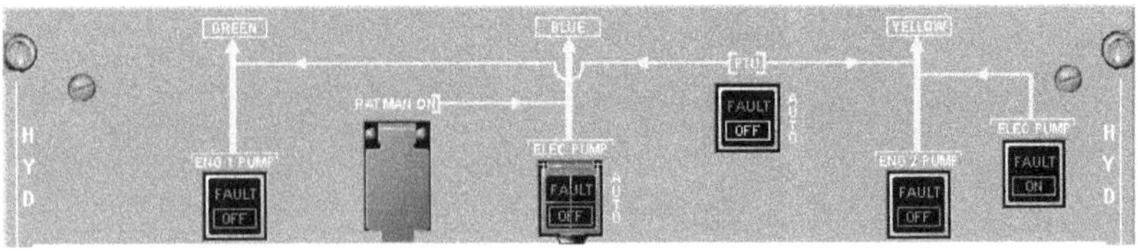

The next ECAM ACTION is set YELLOW ENG 2 PUMP to OFF.

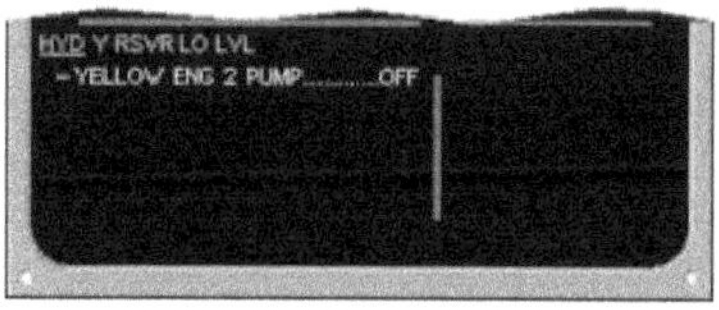

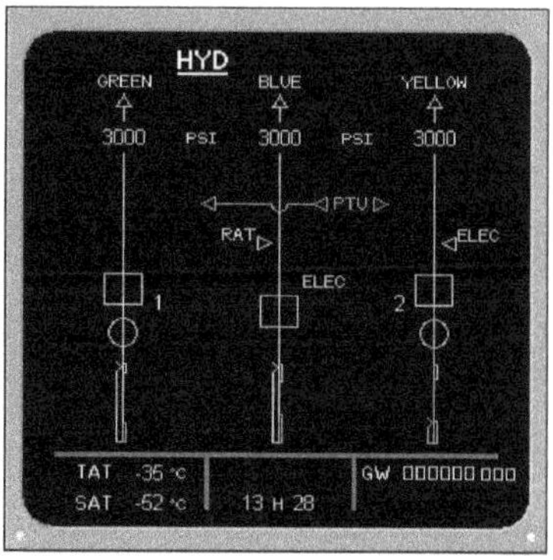

The following indications show that the YELLOW ENG 2 pump has been switched off: Pb sw OFF light, ECAM HYD page, YELLOW ENG 2 pump has changed from green in-line to amber cross-line.

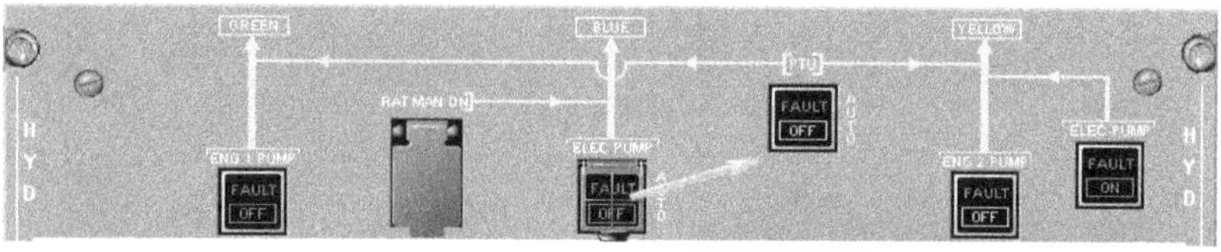

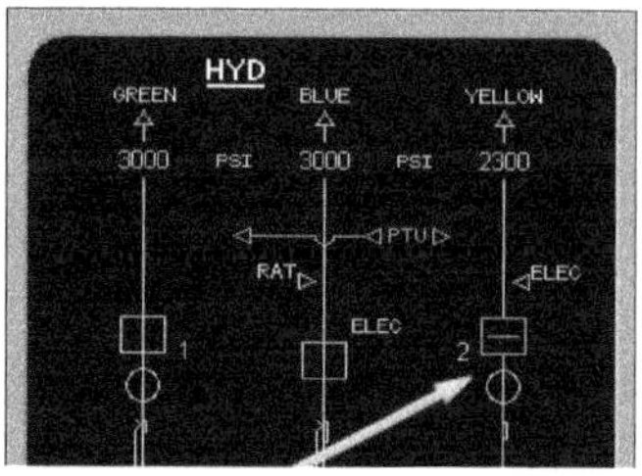

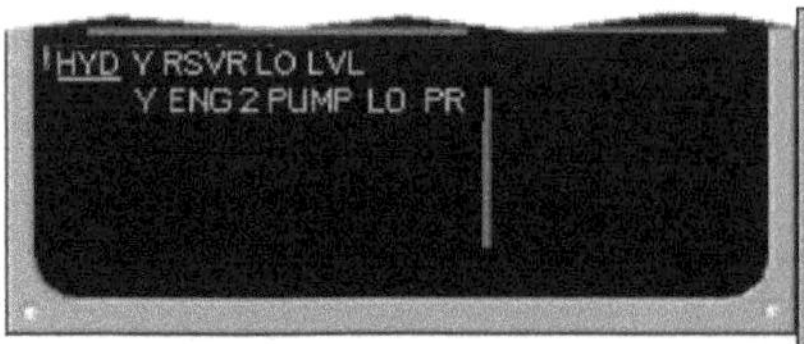

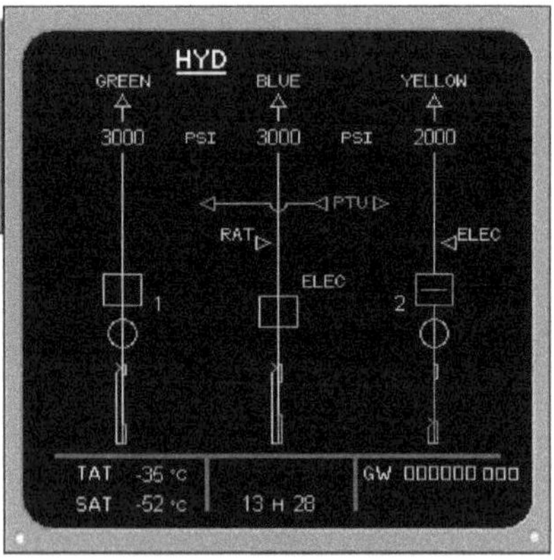

The ECAM Caution, Y ENG 2 PUMP LO PR appears amber on the E/WD to indicate that YELLOW ENG 2 pump is not producing pressure because you switched it off. The YELLOW system pressure continues to drop.

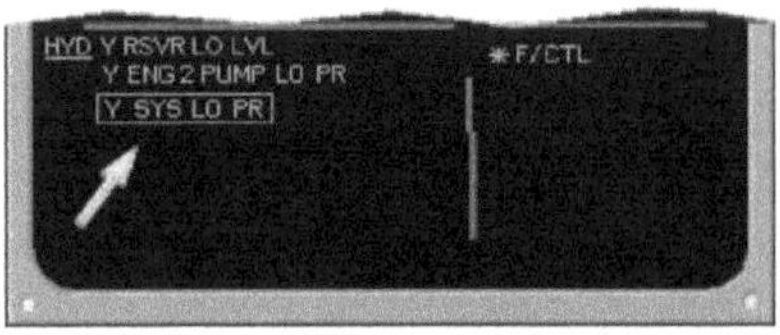

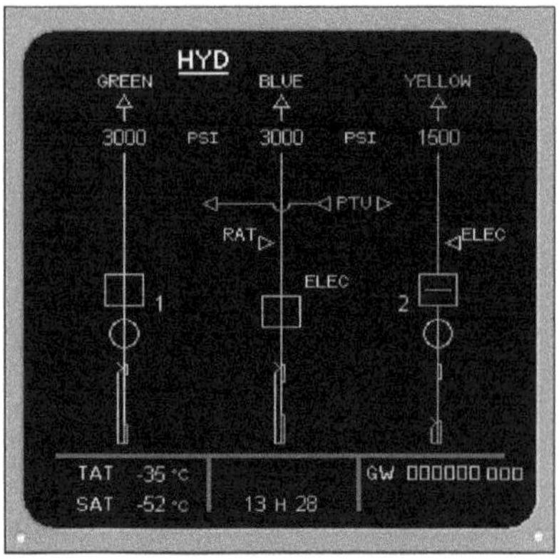

The corresponding ECAM Caution Y SYS LO PR is displayed on the E/WD to indicate that the YELLOW System is no longer pressurized.

At the same time, on the ECAM HYD page, the YELLOW System pressure changes from green to amber. In addition, the YELLOW System indication changes from white to amber, indicating the loss of YELLOW system pressure to the users.

The abnormal procedure is complete, the HYD system is analyzed. Now the STATUS page is displayed for review. ECAM ACTIONS COMPLETED!

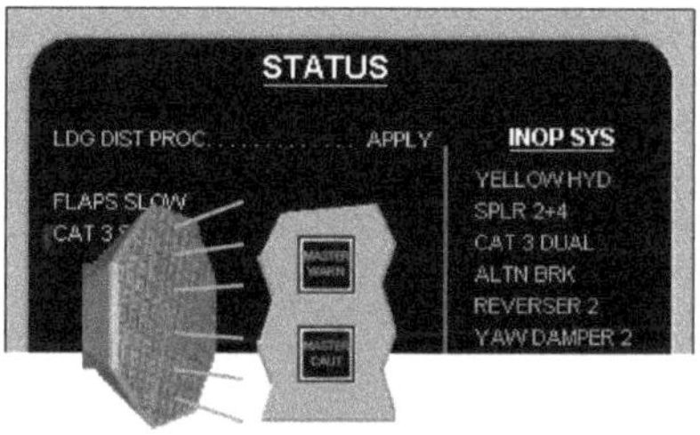

On the E/WD, read the title of the failure. Notice that this is a red WARNING. In addition, LAND ASAP appears in red to indicate the serious nature of the failure. Observe on the ECAM HYD Page, the BLUE Elec Pump indication changes from green in-line to amber LO indicating that the BLUE electric pump is failed. Notice the system pressure is below normal limits. On the Hydraulic Control Panel, FAULT amber is illuminated in the BLUE Elec Pump pb sw which. Let's perform ECAM ACTIONS!. Lift red guard first and activate de RAT.

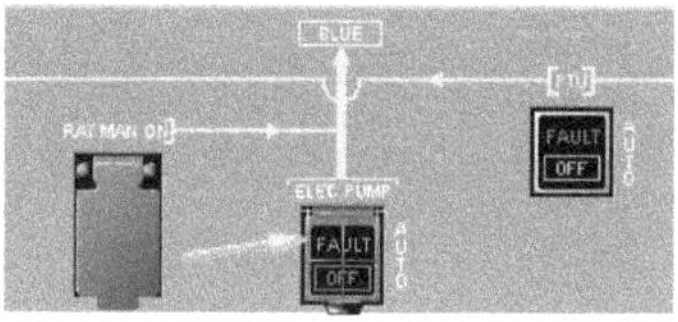

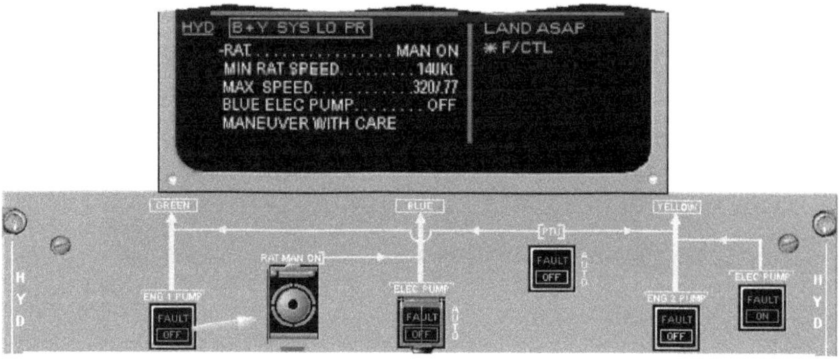

On the ECAM HYD page, observe the RAT symbol has changed from hollow white to solid green indicating that the RAT is extended. Notice that the pressure has increased.

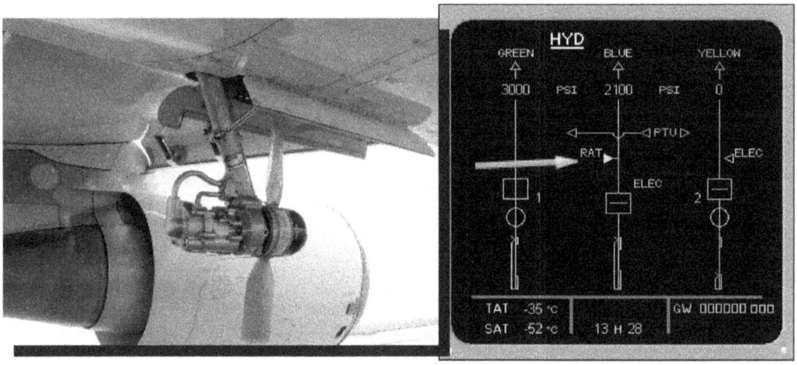

The minimum RAT speed is 140 kt. Note: Once deployed, the RAT can only be restowed by maintenance on the ground. As soon as the RAT has pressurized the BLUE system, the dual Hydraulic System low pressure no longer exists and we are back to a single hydraulic failure situation (Y SYS LO PR).

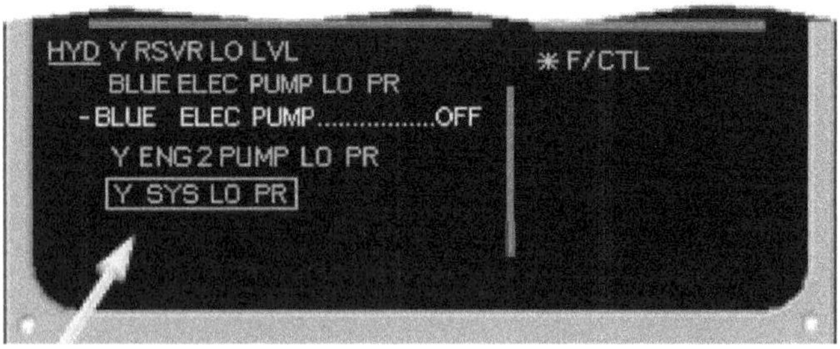

Observe that the BLUE System indication on the ECAM HYD page changes from amber to white. The system is pressurized by the RAT and available for the users.

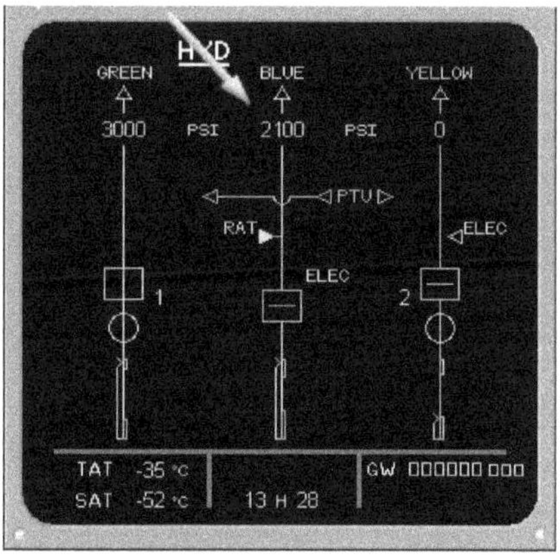

Now, continue ECAM ACTIONS!

BLUE ELEC PUMP ..... OFF

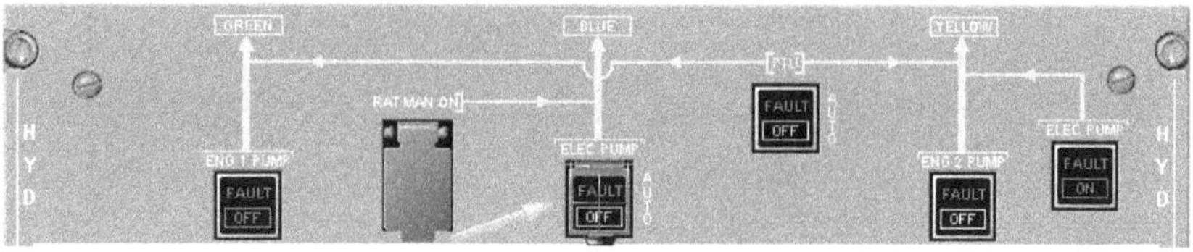

Observe on the System Display, the amber LO indication is replaced by an amber cross-line indication to indicate that the pump is off.

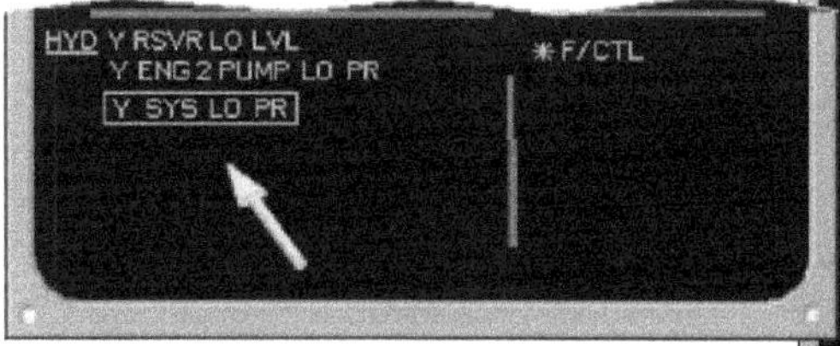

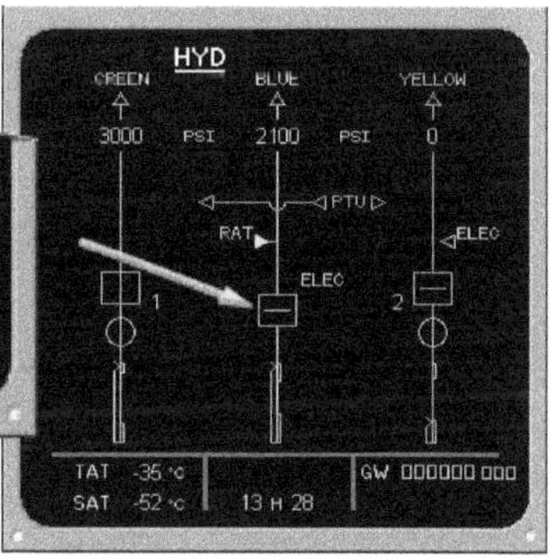

The F/CTL page is automatically displayed to analyze the effects of the primary hydraulic failure. For example, the right elevator is affected by the YELLOW System low pressure.

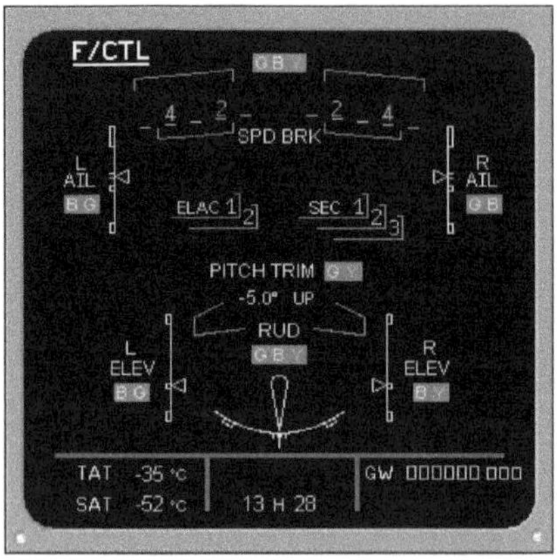

Now, the STATUS page is displayed for review. Notice RAT OUT has reappeared because the F/CTL secondary failure indication, which had priority, has been cleared.

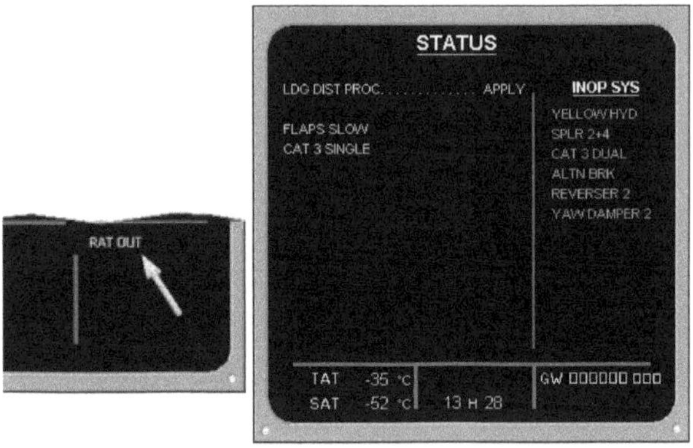

ECAM ACTIONS COMPLETED!

Let's continue. All systems are back to normal. Now let's see other abnormal indications which can be displayed on the ECAM HYD page. If the RAT symbol changes to amber, a fault has been detected in the RAT. If an electric pump overheat is detected:

- The symbol changes from hollow white to a solid amber
- ELEC OVHT in amber is displayed adjacent to it.

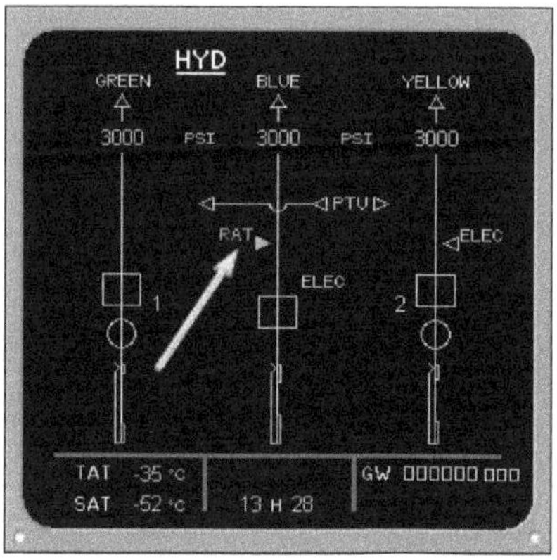

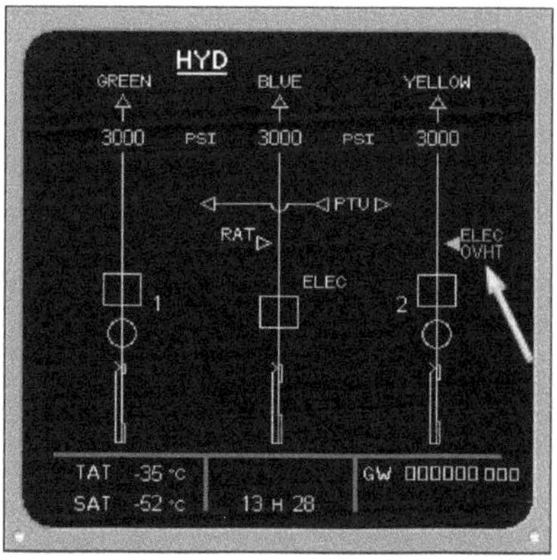

Navigation Abnormal Operation

When we talk about the navigation system, not only means navigation capability, but also aircraft attitude . You are in flight and everything is normal, suddenly the MC is triggered!

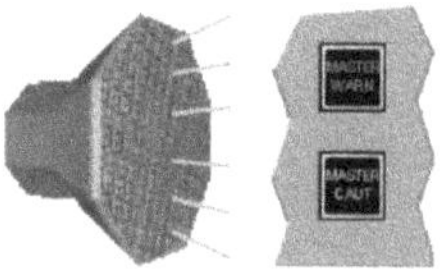

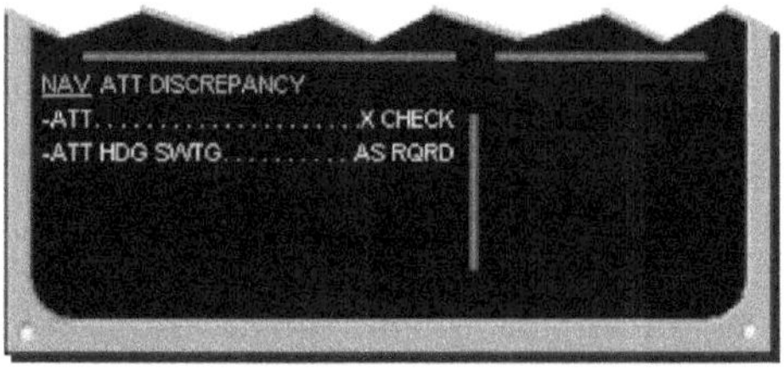

On the E/WD, read the title of the failure. Notice also the "CHECK ATT" on both PFD indicating that there is a difference in the aircraft attitude displayed to the pilots.

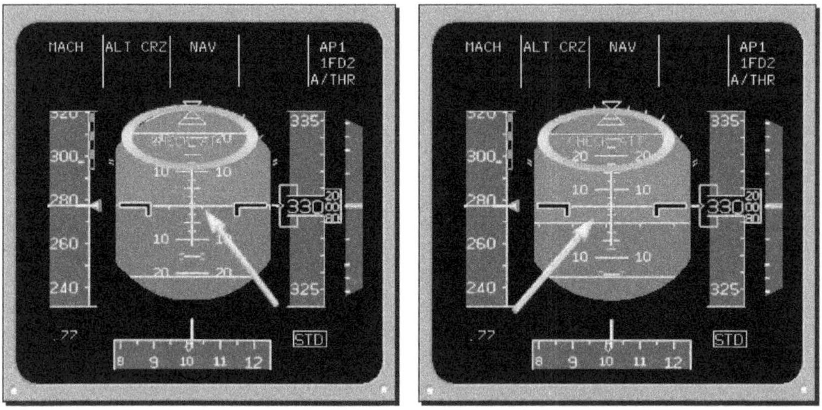

There is no ECAM page associated with the ADIRS, so the CRUISE page remains on the SD. The attitude cross-check demanded by the ECAM procedure is carried out by comparing the PFD attitudes with that of the standby horizon. In this example, the F/O PFD, which shows a different attitude to that given by the captain's PFD and the standby horizon, is likely to be the one at fault.

Having identified which PFD is giving the incorrect attitude display, the ATT HDG switch should be selected to that side. In this case, the switch must be selected to F/O 3 position. Let's see what happen...

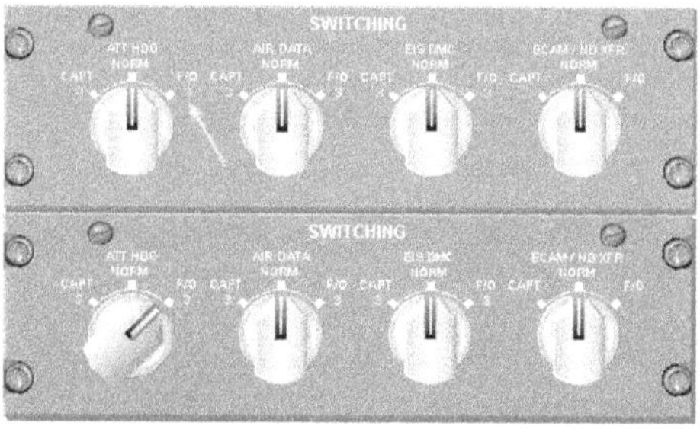

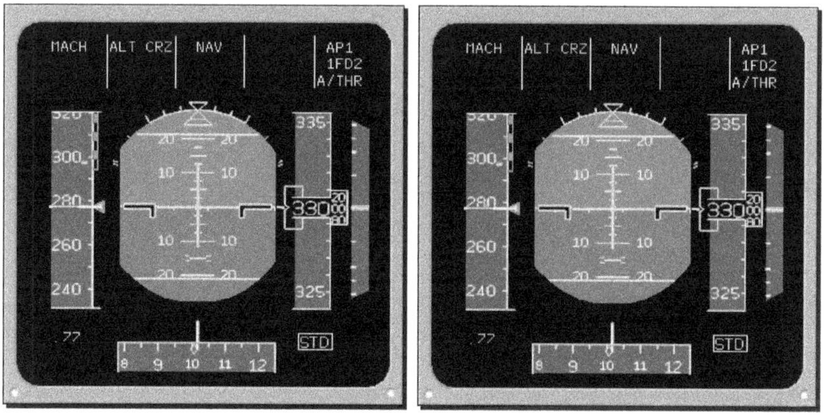

IR 3 is now supplying the F/O side. The aircraft attitude on the F/O PFD is back to normal. "CHECK ATT" indications have disappeared. On the secondary failure section will appear the message SWTCHG PNL. ECAM ACTIONS COMPLETED!

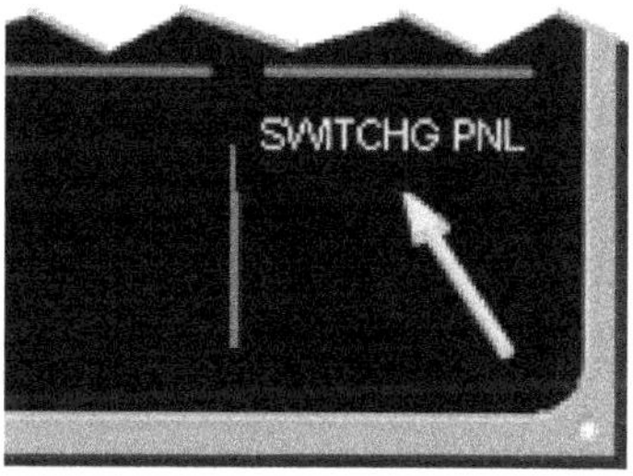

We will continue with an ADR 1 fault which normally supplies ATC 1. Let's assume, you are the pilot non-flying. You are in flight, in cruise and everything is normal. Suddenly the MC is triggered! On the E/WD, read the title of the failure. Notice on the PFD, the speed, altitude, mach and barometric indications are lost. The procedure requests you to switch the AIR DATA sw to the CAPT position. Let's do it.

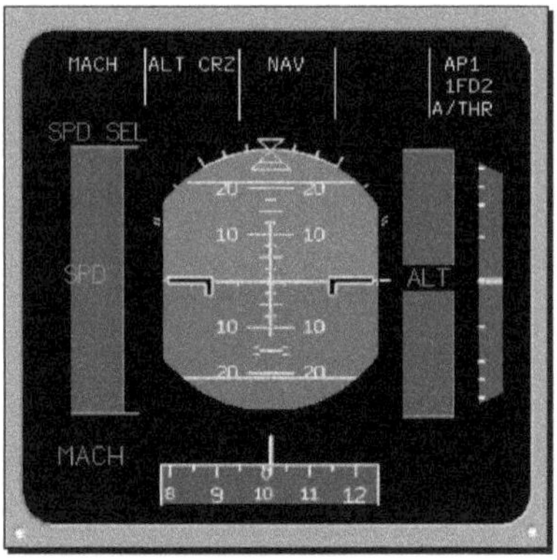

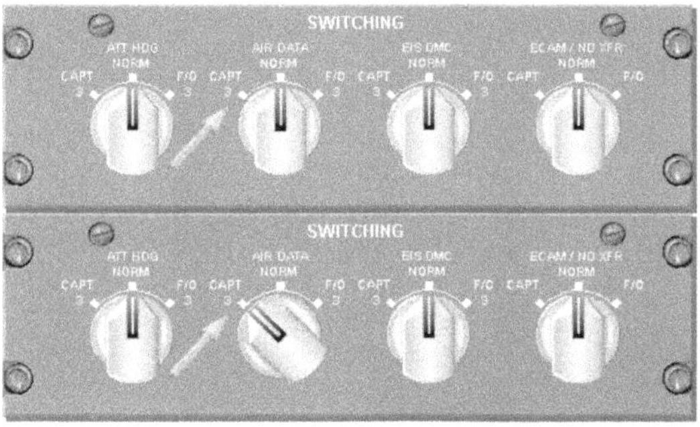

By switching the AIR DATA sw to CAPT 3, ADR 3 supplies the Captain's PFD, ND and the DDRMI instead of the faulty ADR 1. Continue ECAM ACTIONS….

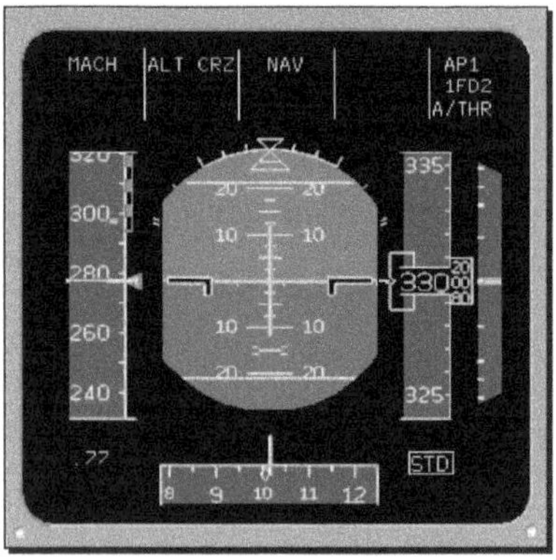

ADR 1 has to be switched off. ADR 1 is now off which has led to a GPWS fault.

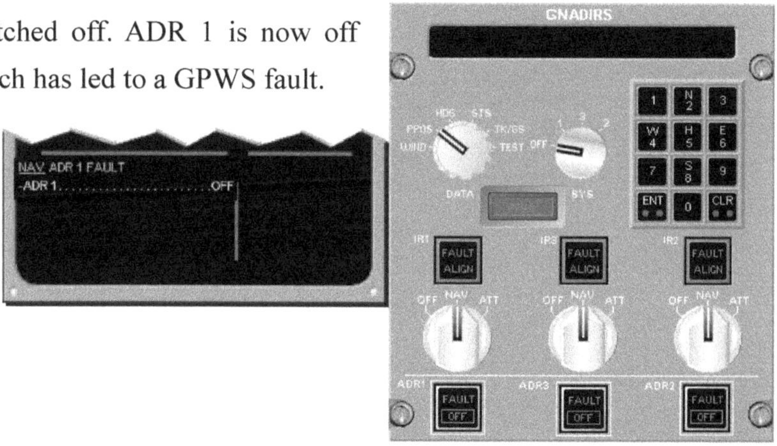

GPWS has to be switched off.

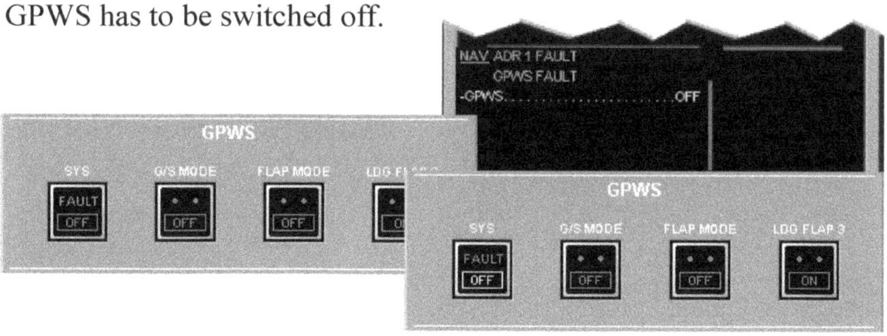

The STATUS page is displayed for review. A green message advises us that the approach capability is degraded to CAT 3 SINGLE ONLY. You can see in the INOP SYS column that ADR 1 and CAT 3 DUAL are now inoperative. The GPWS is also unavailable due to ADR 1 fault.

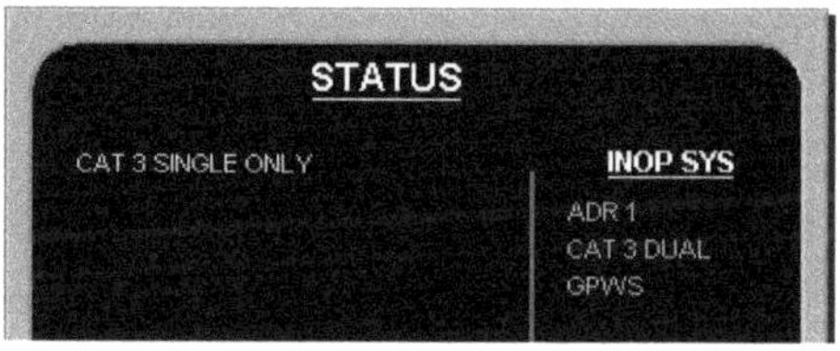

Now let's see now a demonstration of an in-flight RADIO ALTIMETER abnormality. You are the pilot non-flying. You are in descent everything is back to normal and suddenly...

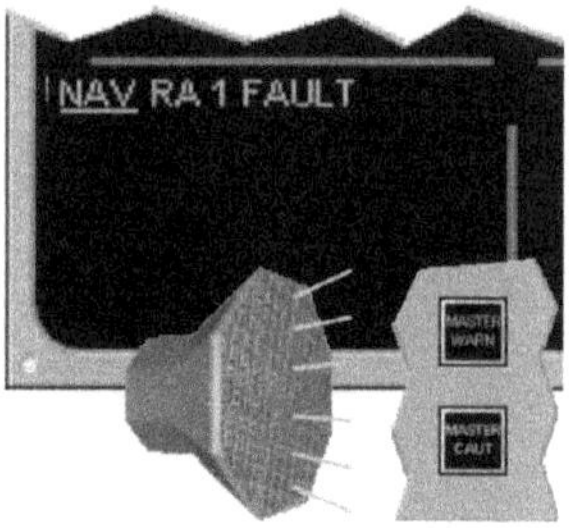

On the E/WD, read the title of the failure. As there is no associated procedure, after review and confirmation from the PF. The STATUS page is displayed, showing inoperative systems and aircraft information. With only one RA, CAT III is unavailable, and the failure of RA 1 causes the loss of the GPWS.

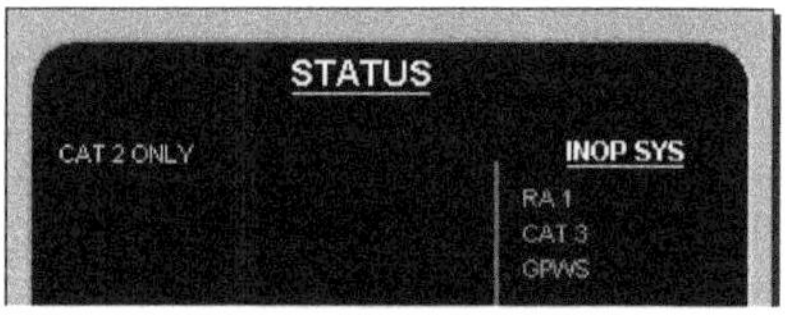

You are flying an approach manually, with RA 1 still faulty and suddenly

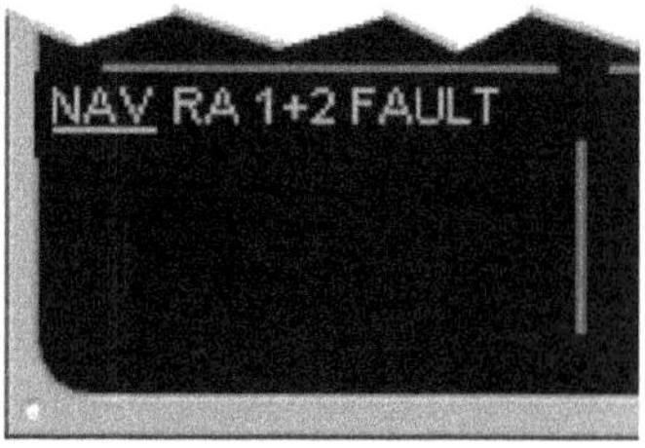

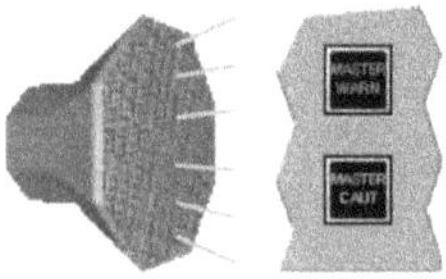

The STATUS page is displayed, showing inoperative systems and aircraft information. With both RA faulty, both AP are lost in

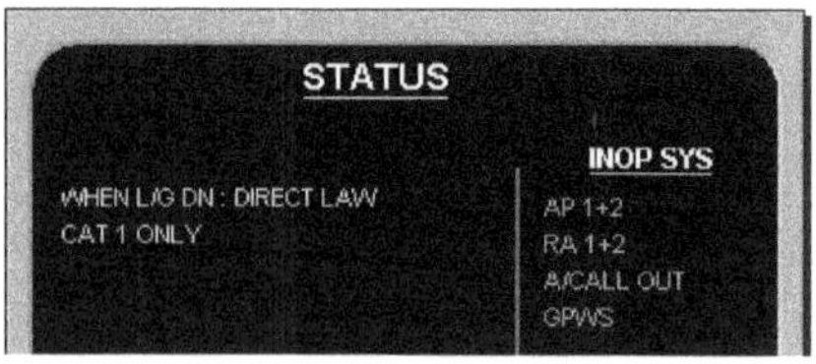

approach, as well as the auto call-outs.. ECAM ACTIONS COMPLETED!

You will now study how to navigate using the standby navigation keys. Let's assume, you are the pilot non-flying , in cruise and this is one of those days... you have just lost both FMGC.

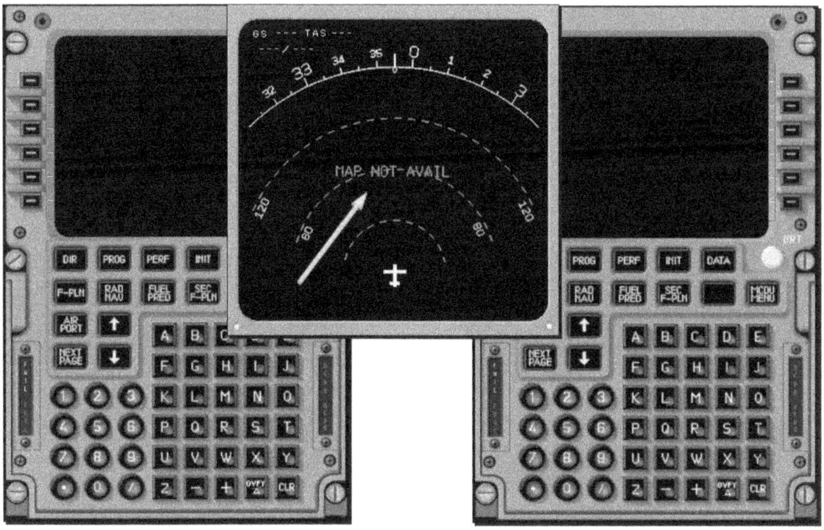

Blank MCDU screens with amber FAIL annunciator lights, and the red map on the ND indicate that a double FMGC failure has occurred. Here, we will focus on the navigation, so we will consider that all the actions concerning the other systems have been performed.

As the MCDU are no longer available for navigation and nav aid tuning, the only navigation capability remaining is radio aid raw data.For this we need to use the standby nav aid tuning capability of the RMP.

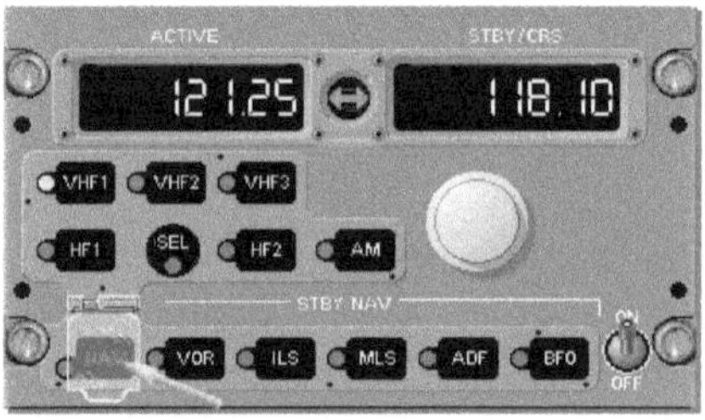

The green NAV light illuminates to indicate that RMP 1 is now in NAV mode.

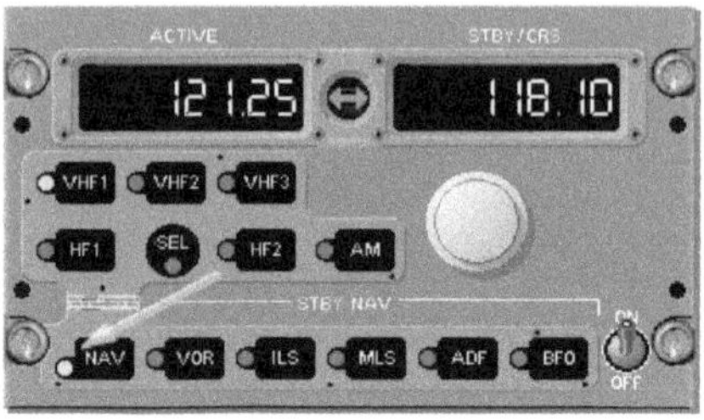

For nav aid tuning, RMP 1 STBY NAV keys are associated with VOR/DME 1 and ADF 1 while RMP 2 keys are associated with VOR/DME 2 and ADF 2.

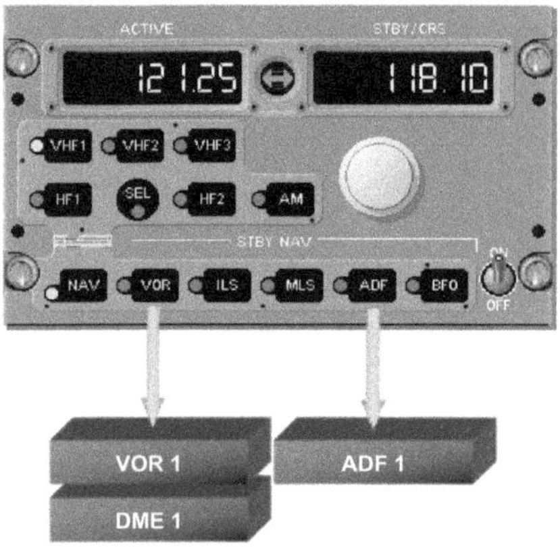

The STBY NAV ILS key is illuminated. The ACTIVE window is now showing up the last memorized ILS frequency while the STBY/

CRS window is displaying the last memorized ILS course. Notice also that the selected radio light is now extinguished.

To use the RMP for communication frequency change, you must deselect the STBY NAV mode by pressing the NAV key. Any nav aids using STBY NAV mode will remain tuned. We will now review some indications we did not have the opportunity to see when performing the previous procedures. We will see first an ATC unit failure. The code display goes blank and the ATC FAIL light illuminates when the selected transponder is failed, in this example, transponder 1. System operation may be recovered by selecting the other transponder, the number 2.

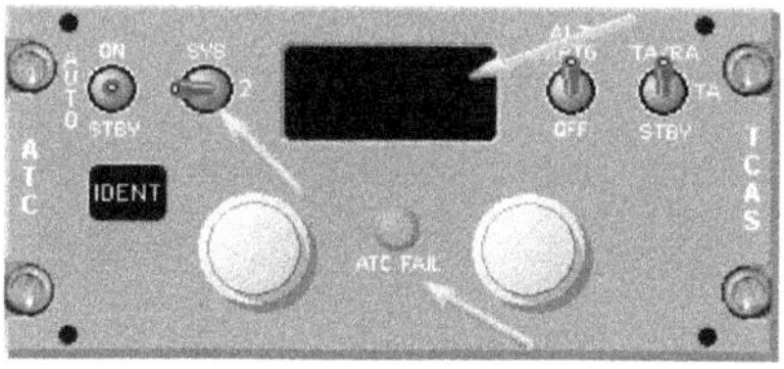

With the selection of the second transponder, the code reappears in the display and the ATC FAIL light extinguishes. The ATC/TCAS unit is again fully operational. We will continue with the weather radar failure messages. In case of any failure, a message appears on the ND : amber for a radar performance degradation, red when no image is available.

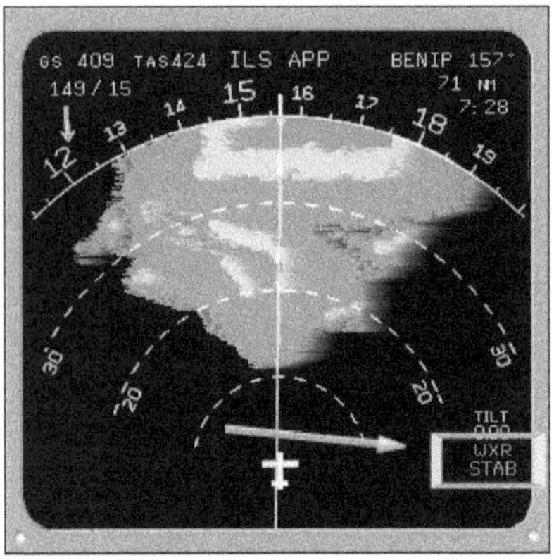

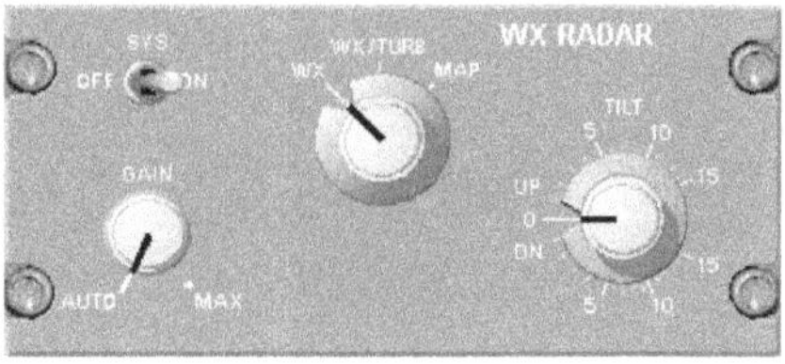

The next faulty system is the TCAS. In case of TCAS failure, a TCAS fault is displayed on the E/WD and TCAS red flags appear on both PFD and ND. Note that there are neither warning light nor aural warning.

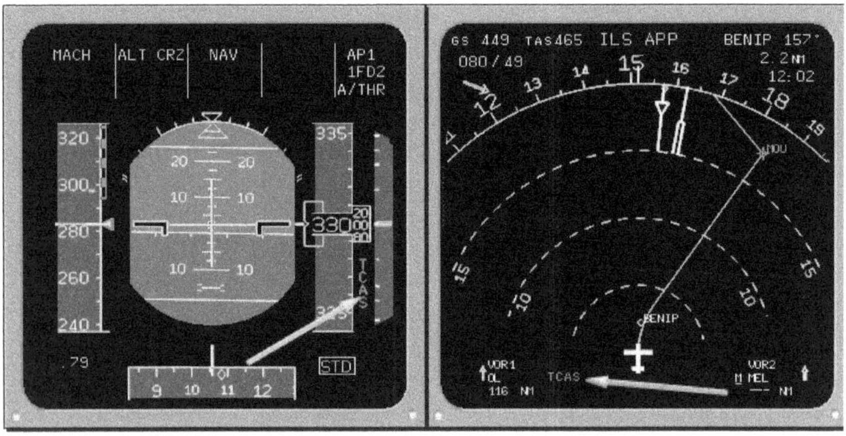

Chapter 3

Abnormal operations III

Aviation Bookstore
aviation step by step

NOTE: information not valid for real flights. Concepts explained as examples for your study and entertainment. For real flights see the original manuals of the manufacturer AIRBUS.

Power Plant Abnormal Operation

This is a demonstration of an in-flight loss of EPR mode. It will be replaced by N1. This is reversion to rated N1 MODE. If additional EPR computed parameters are not available, the new degraded mode is called unrated N1 MODE. You are in flight and everything is normal, suddenly the MC is triggered!

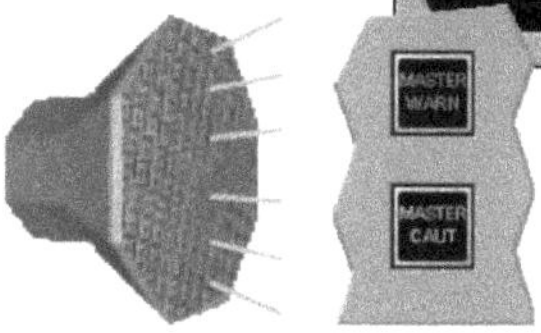

The amber warning and associated checklist are displayed on the E/WD. Read the title of the failure. When EPR is lost the autothrust is disconnected. The EPR indicator is amber and the value is crossed amber showing that EPR 1 MODE is lost. The actual N1 is shown both numerically and by the needle in green. The needle and the numeric value pulse amber when N1 is greater than the N1 rating limit. They pulse red when N1 is greater than 100 %.

The white circle indicates the thrust lever position. This white circle is not displayed in unrated mode. The amber index indicates the full forward position of the thrust lever (MAX N1). MAX N1 is not displayed in unrated mode. Let's perform ECAM ACTIONS! Set EGN 1 N1 MODE...... ON

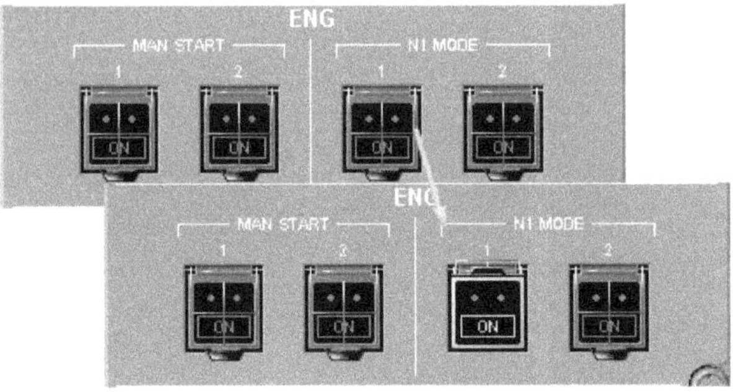

On the ENG panel, the engine 1 N1 MODE pb switch changes to ON blue. Now, continue ECAM...

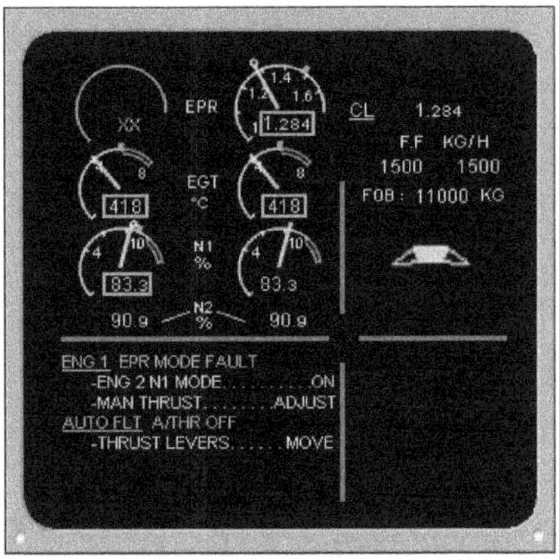

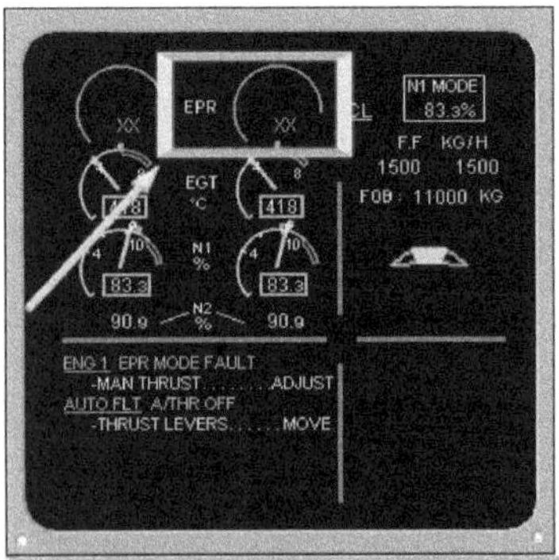

Engine 2 EPR indication turns amber indicating that the EPR MODE is lost. Next action is to adjust the thrust manually depending on the request. Assume this has been done. ***ECAM actions complete!***

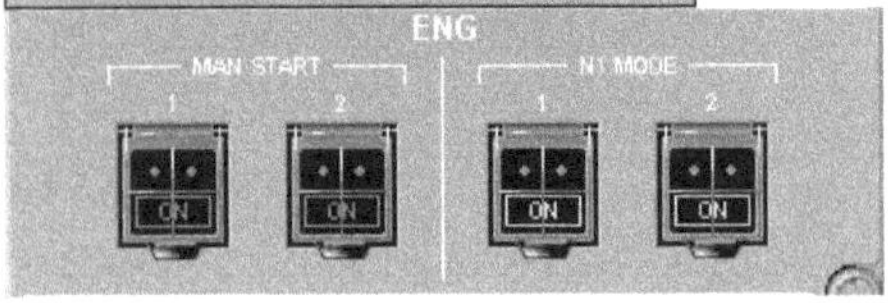

Let's look briefly at some other engine abnormal indications. The amber warning is displayed on the E/WD. Read the title of the failure. On the system display, the ECAM ENGINE page is automatically displayed. The oil filter indication is displayed and the corresponding oil filter is indicated by CLOG in amber.

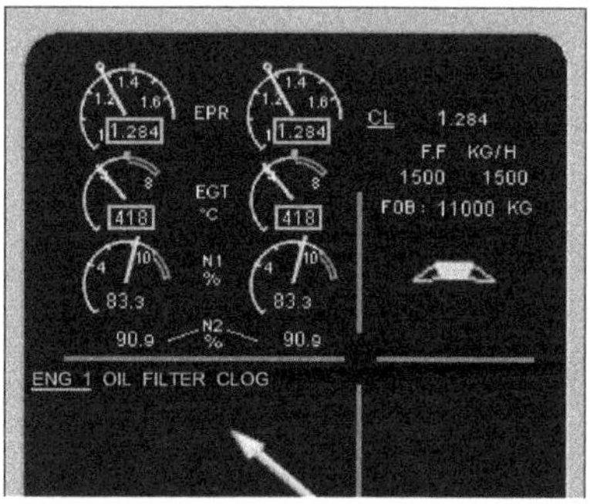

The same kind of message can occur with the fuel system. The amber warning is displayed on the E/WD. On the system display, the ECAM ENGINE page is automatically displayed. The fuel filter indication is displayed and the corresponding fuel filter is indicated by CLOG in amber.

Now, this is a demonstration of an abnormal engine start. To do that, let's start the engine 2 in normally sequence. Notice after changing the mode selector from NORM to IGN START, the engine parameter on ECAM will change.

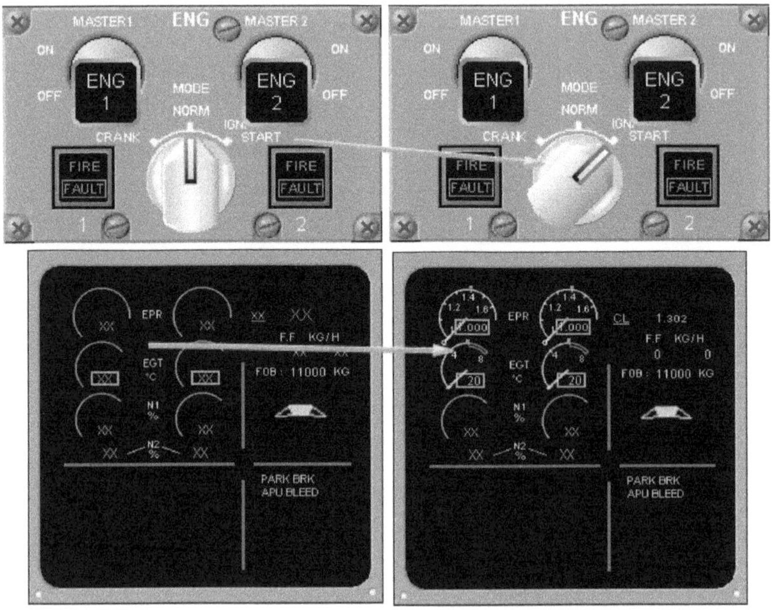

The next step in the normal engine start is to set ENG 2 MASTER to ON. Notice the ignition channel is open and the N2 parameter is increasing (24.0).

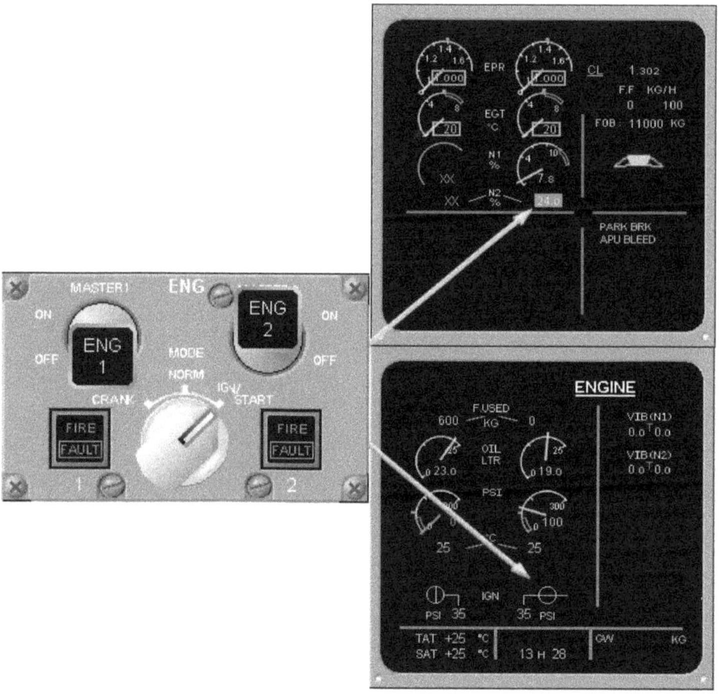

The normal sequence is carrying out and suddenly, the MC alarm is triggered..As there is no ignition, the caution is activated and the amber warning and associated checklist are displayed on the E/WD. Read the title of the failure.

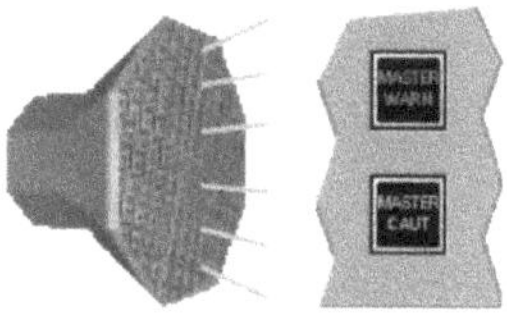

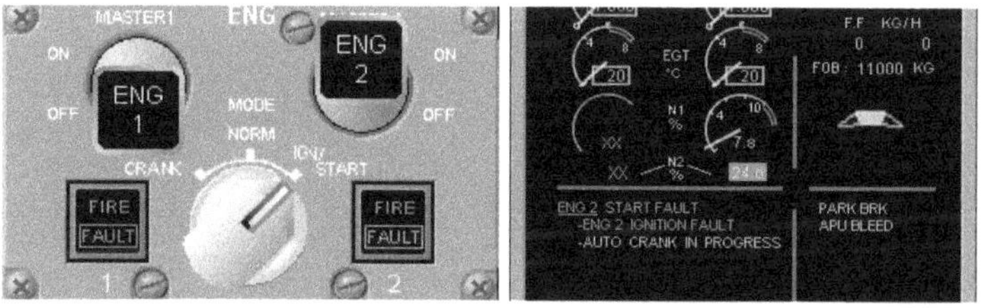

On the ENG panel, the FAULT light illuminates amber indicating that the automatic start is aborted. Automatically the FADEC shuts off the fuel, turns off the ignition and closes the start valve. The message AUTO CRANK IN PROGRESS is displayed on the E/WD indicating that the FADEC automatically performs a dry crank. After seconds, the auto crank cycle is complete. The message on the E/WD asks you to confirm the automatic start abort. The PF will ask you to perform the ECAM actions.

When the engine 2 master switch is off, the message disappears on the E/WD and the fault light is switched off. We will assume that the procedure is complete.

Now let's see another engine abnormal operation.

You are the PNF. Everything is normal. We are in cruise and we have displayed the ECAM ENGINE page for you. The engine 1 oil pressure indicator pulses amber indicating that the oil pressure is below the normal range. Suddenly the MW alarm is triggered!

The red warning and associated checklist are displayed on the E/WD. On the system display, the ECAM ENGINE page is automatically displayed. The engine 1 oil pressure indicator is displayed in red indicating that the oil pressure is too low. The procedure is to shut down the faulty engine and that's all. We will assume that the procedure is complete.

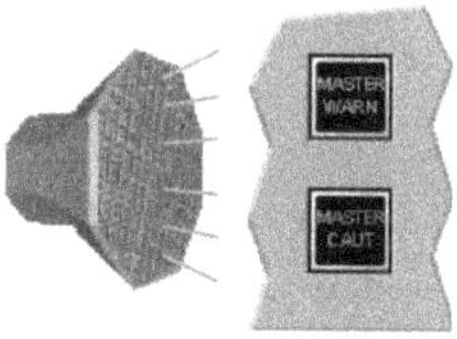

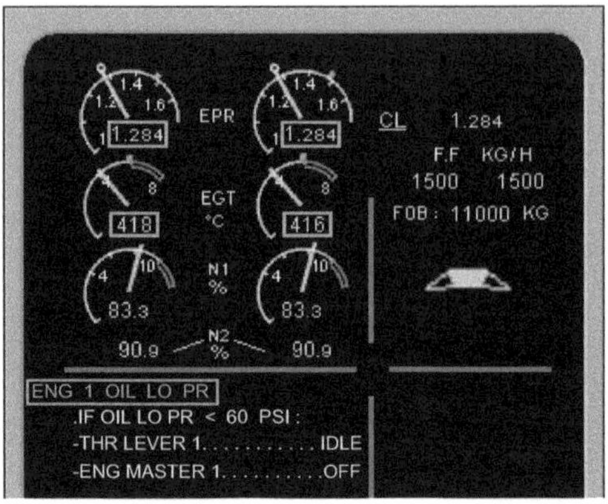

Now let's see another engine abnormal operation. You are on ground, everything is normal. Suddenly the MC alarm is triggered! The amber warning and associated checklist are displayed on the E/WD. Notice that we are operating in very cold weather. Observe the oil temperature is still displayed in green.

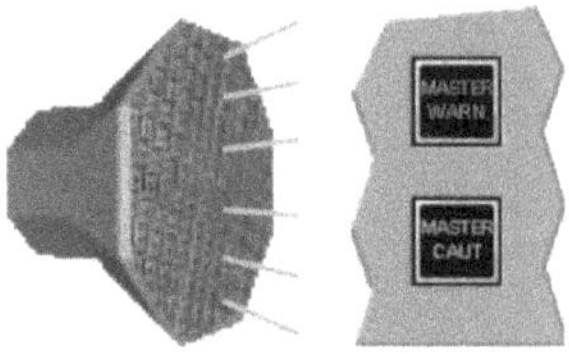

The procedure instructs you to delay take off. The engine should be warmed up prior to take off. Observe that if the oil temperature is above the normal range, the oil temperature indication pulses green.

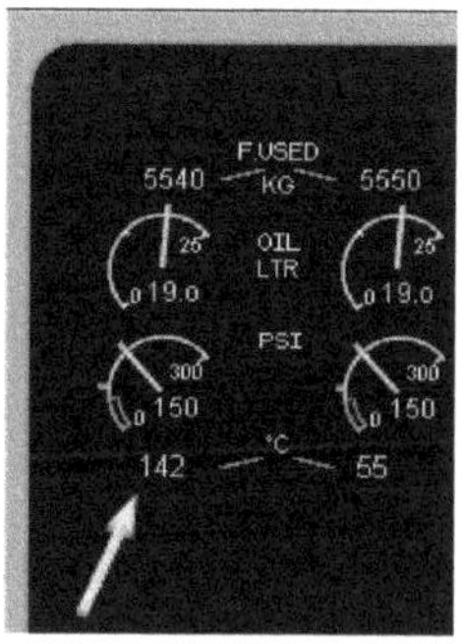

Now, observe that if the oil temperature continues to increase, an ECAM caution is activated above 165 degrees Celsius and the oil temperature indication becomes amber. In the secondary failure section the message LANAD ASAP amber appears. The plane has to land as soon as possible. Notice that it's not a RED LAND ASAP!

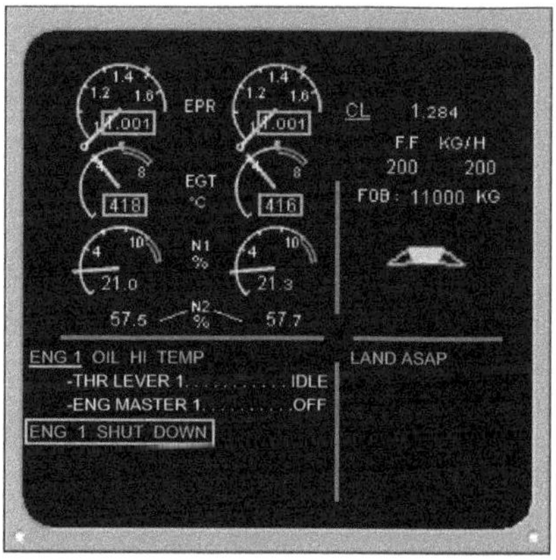

We will now look at some examples of other abnormal indications which can be displayed on the ECAM ENGINE page. The nacelle temperature is displayed as soon as one nacelle temperature is above the normal range. When displayed, only the nacelle temperature which is over the normal limit, pulses green.

During the start sequence with the start valve opened, if the upstream bleed pressure is below the normal range, the indication becomes amber. Note: this may happen during start but the pressure usually returns to normal. If the pressure remains low, await ECAM instructions.

Let's continue. We will see a demonstration of an engine EGT over limit in cruise. You are the PNF. We are in flight and everything is normal. On engine 2, EGT value is increasing.

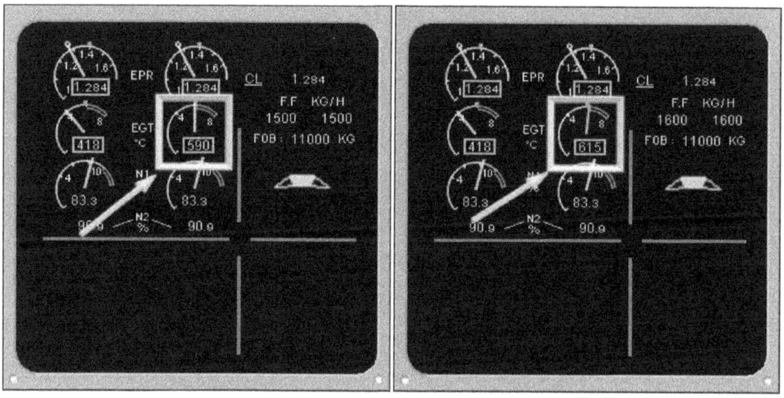

You heard and saw the caution. The amber caution and associated checklist are displayed on the E/WD. The temperature has continued increasing. On the EGT indicator, the indications have turned red and flash. The only ECAM action is to move the thrust lever. We will do this for you.

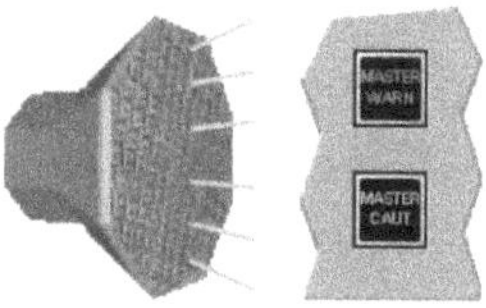

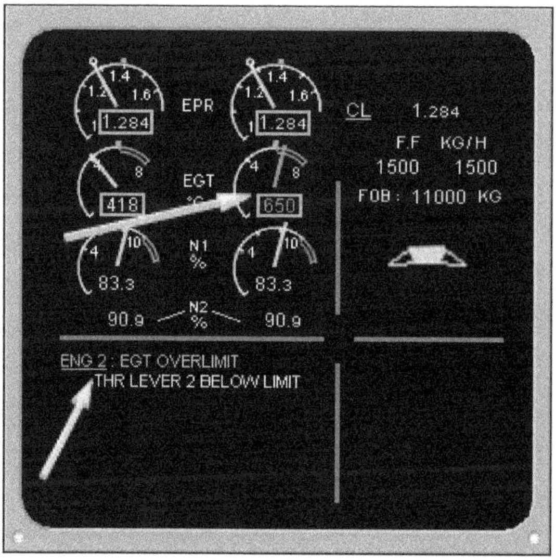

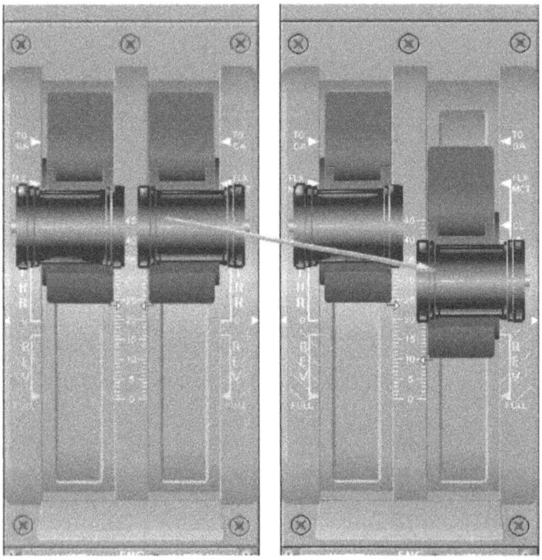

The EGT engine 2 value indication changes from red to green. ECAM actions completed.

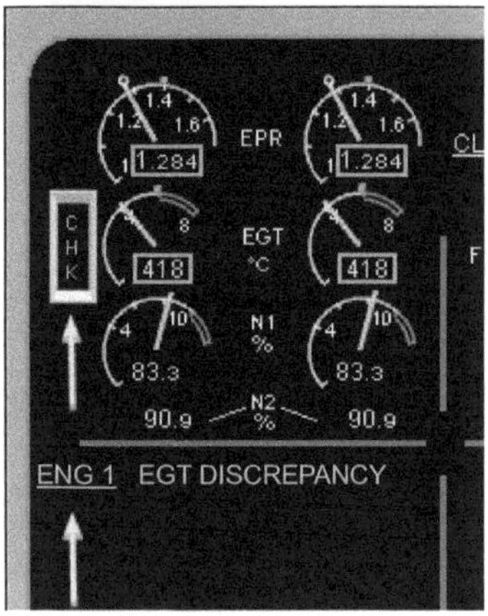

We will see now another engine abnormal indication on the E/WD. We are in flight and everything is normal. Suddenly the MC alarm again! An amber CHECK message is displayed beside the engine 1 EGT indicator, indicating that there is a discrepancy between the value displayed on the ECAM and the real value.

Ice and Rain Abnormal Operation

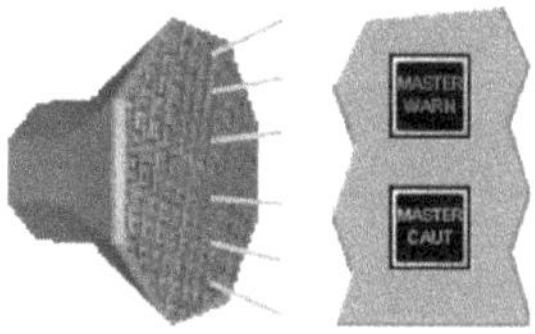

This is a demonstration of an in-flight failure of the left windshield heating system. You are currently flying at your cruise level, everything is normal, suddenly the MC is triggered!

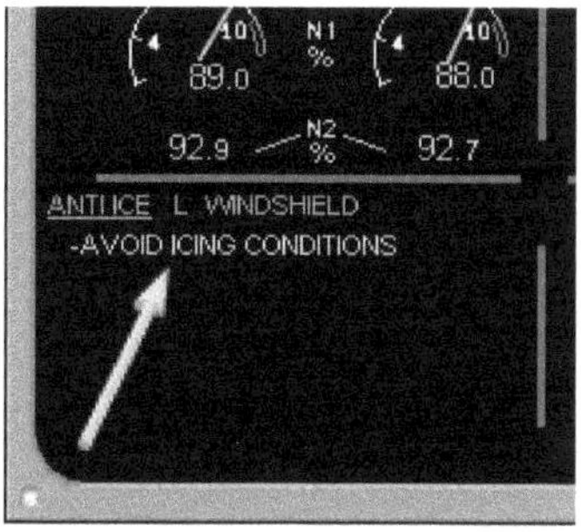

On the E/WD, read the title of the failure. In this example, the blue line advises you to avoid icing conditions. Notice that no system page is called. Having noted the advice and after review and confirmation from the PF…

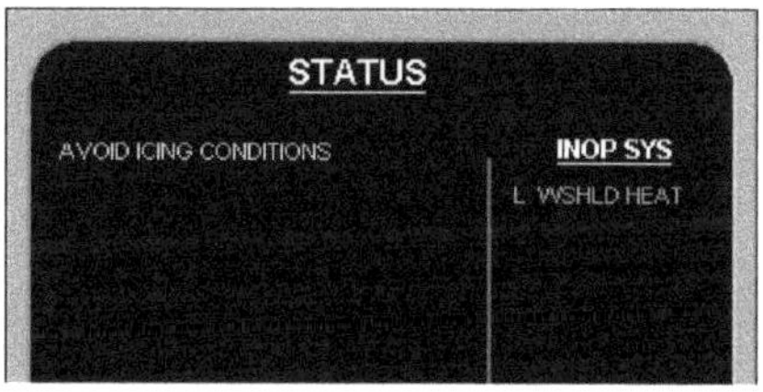

The STATUS page is now displayed and reminds you to avoid icing conditions. The inoperative system is the left windshield heating. ECAM ACTIONS COMPLETED!

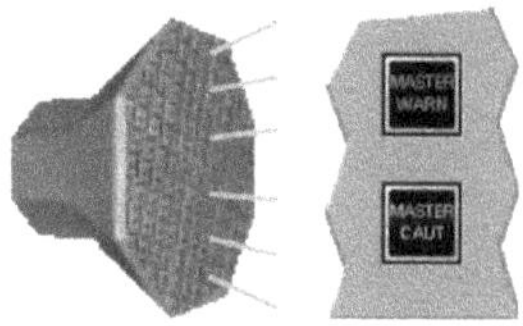

Let's continue with an in-flight Captain's pitot heating failure. You are still in cruise, everything is back to normal, suddenly the MC is triggered!

On the E/WD, read the title of the failure. This time, the Captain's pitot probe heating is not operating. This could lead to ice build up on the Captain's pitot causing incorrect speed information. To ensure valid information to the Captain's instruments, the air data source should be changed. The PF will ask you to perform the ECAM actions.

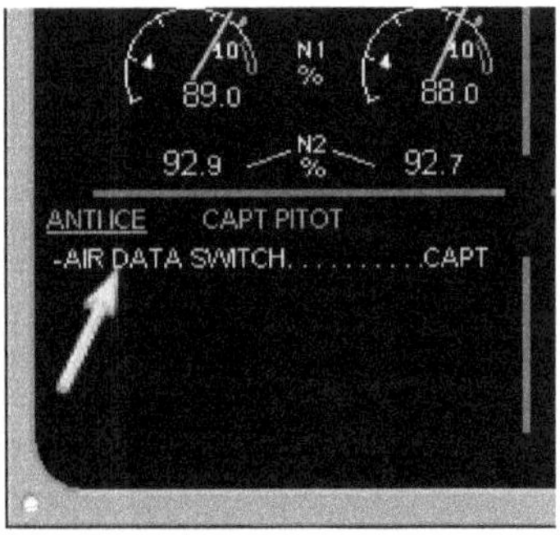

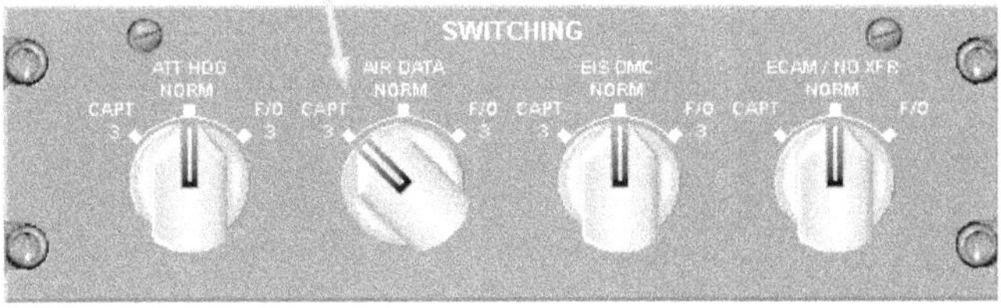

By selecting CAPT 3 on the AIR DATA selector, ADR 3 replaces ADR 1 to provide valid data to the Captain's instruments. Note: AIR DATA switching should not be selected to CAPT 3 if ADR 3 is not available.

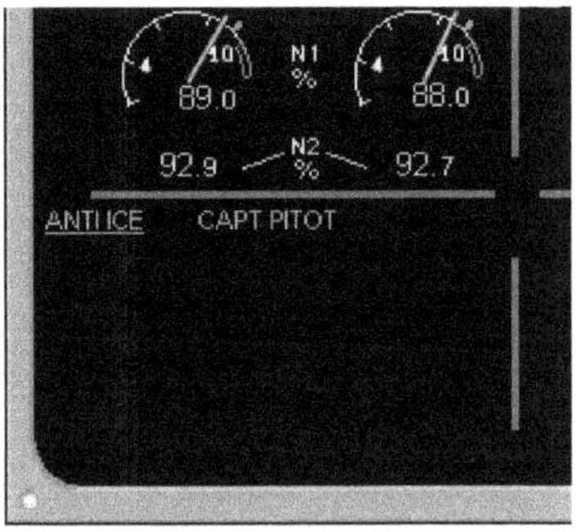

On the E/WD, the memo message “SWITCHING PNL “ is displayed to remind you that the Captain’s ADR has been switched. The STATUS page is displayed. The inoperative system is the Captain’s pitot heating. ECAM ACTIONS COMPLETED!

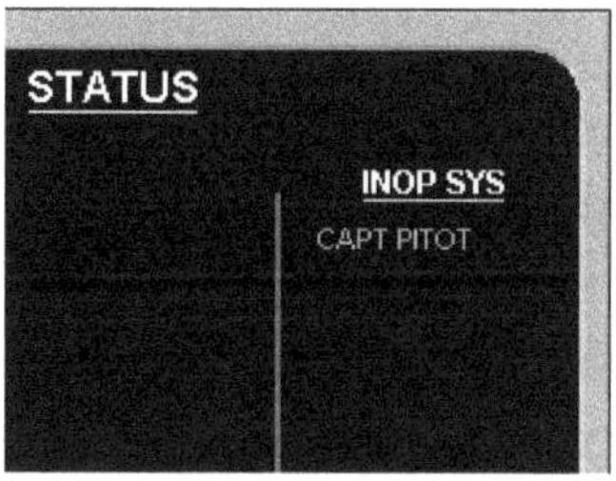

Suddenly the MC is triggered again. On the E/WD, read the title of the failure. The ECAM caution “WING A.ICE SYS FAULT” appears meaning that at least one wing anti-ice valve has not opened after switching the system on. Notice the ECAM BLEED page is called automatically. Before applying the ECAM actions, you should clear the first message.

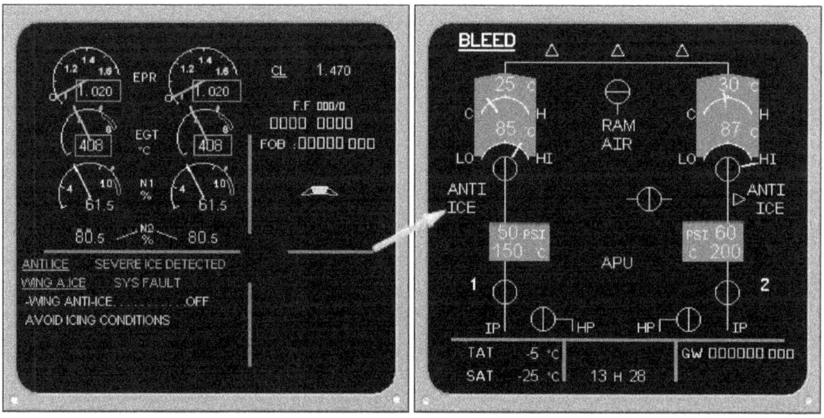

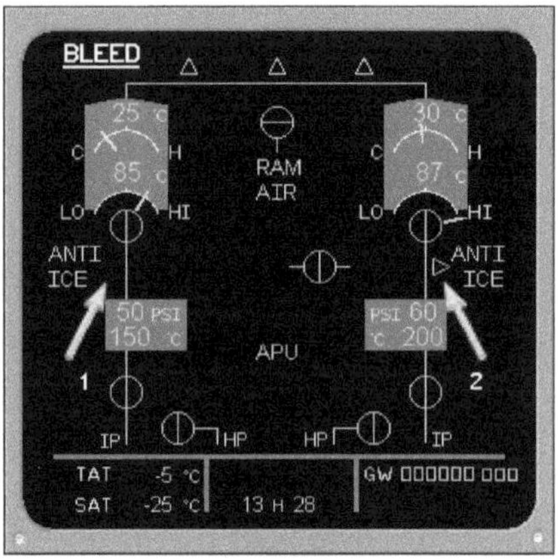

On the ECAM BLEED page, there is no green arrow on the left side, meaning that the left wing anti-ice valve is not open. Observe that the FAULT light in the WING anti-ice pb sw remains illuminated. The PF will ask you to perform the ECAM actions.

On the system page, all wing anti-ice indications have disappeared, meaning that wing anti icing is no longer operating. The ECAM advises you to avoid icing conditions.

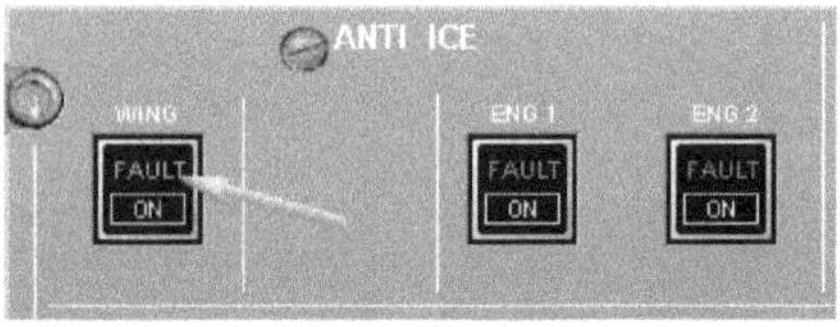

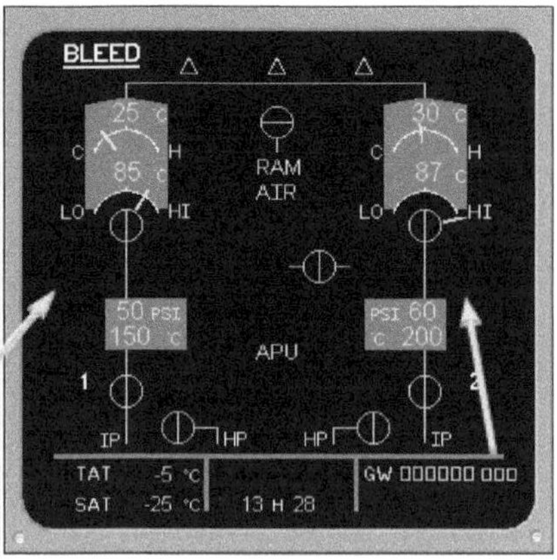

The STATUS page is called. The message "AVOID ICING CONDITIONS" is displayed. The inoperative system is the wing anti-ice. ECAM ACTIONS COMPLETED!

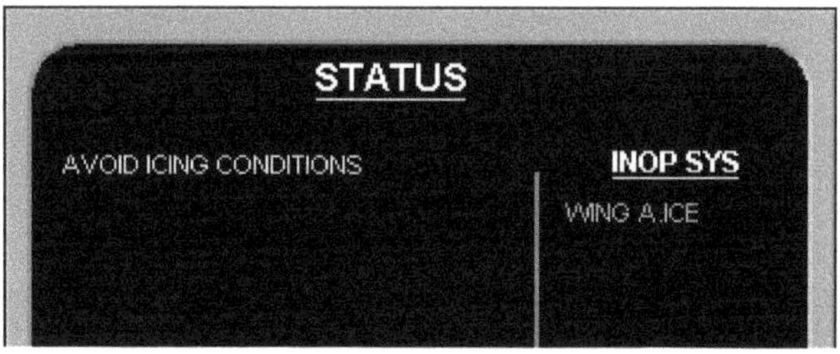

Fire Protection Abnormal Operation

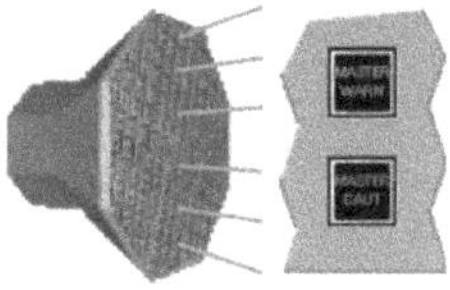

Let's continue with one of the most important system protection. You are in flight and everything is normal, suddenly the MW **ENG 1 FIRE** is triggered!

On the E/WD, the message ENG 1 FIRE with the associated procedure is displayed. The illuminated FIRE pb sw on the FIRE panel and the FIRE light on the ENG panel confirm this indication. The SD shows the corresponding page with the nacelle temperature pulsing.

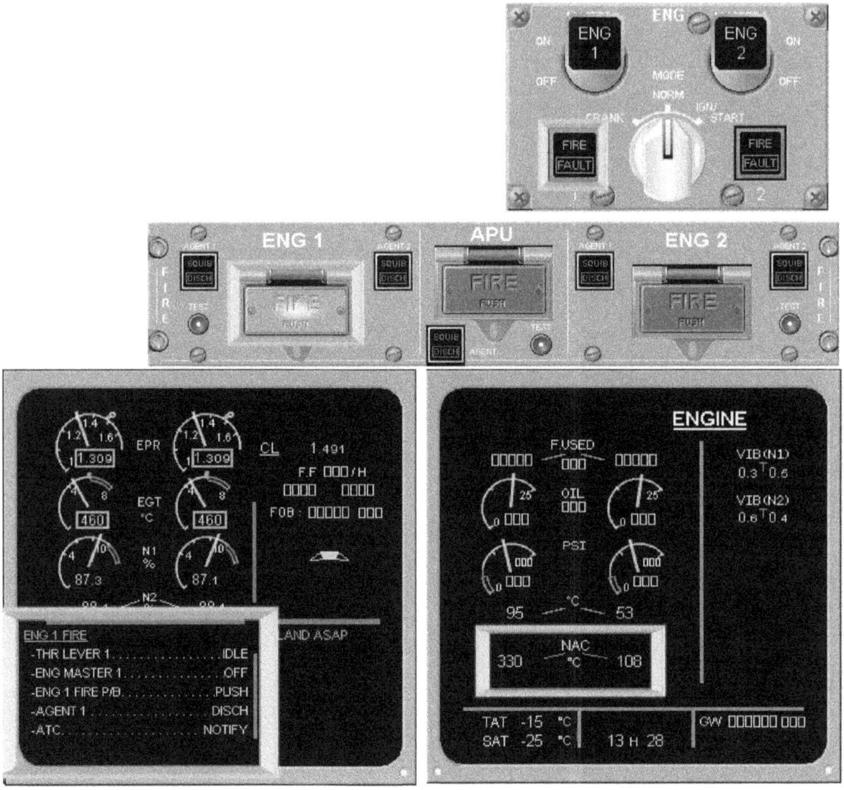

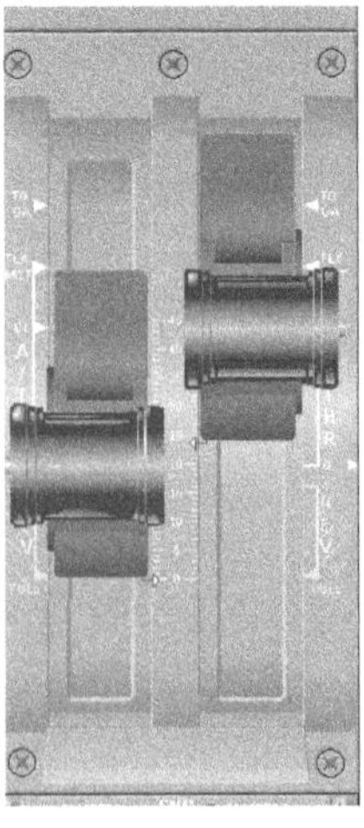

The PF maintains control of the aircraft and asks you, as PNF, to perform the ECAM actions. Engine 1 is spooling down, engine 2 is accelerating to compensate for the thrust loss. You now have exactly the same situation as for an engine failure, so the PF sets the thrust lever 2 to MCT. Let's continue ECAM ACTIONS...

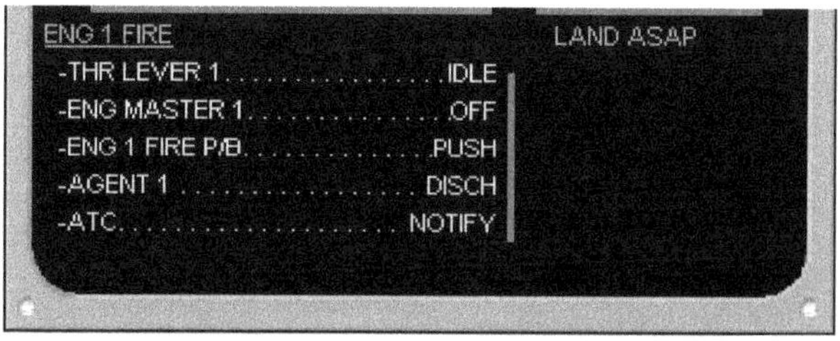

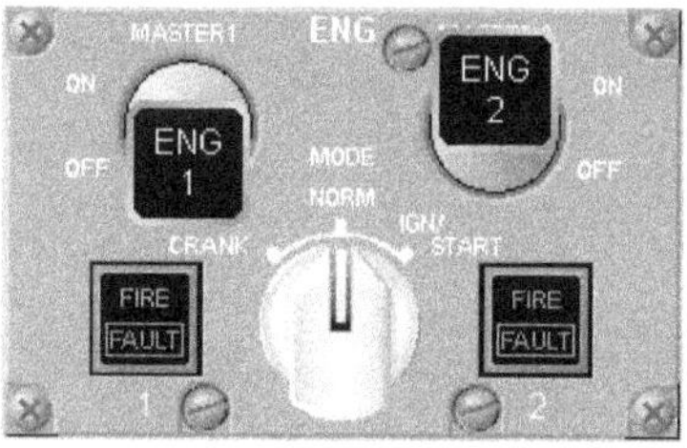

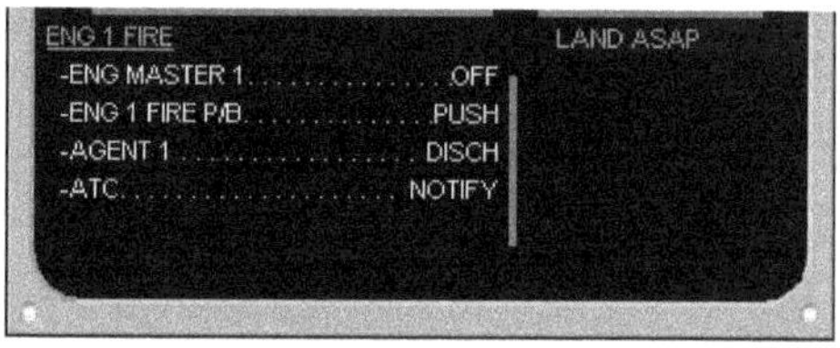

After having switched ENG 1 off, the after ENG 1 SHUT DOWN procedure is shown on the E/WD. Since the IP and HP valves close and the generator is no longer operating, AIR BLEED and ELEC are shown as secondary failures. After the engine has been switched off, you can now go ahead and take care of the fire itself. To be able to push the ENG 1 FIRE pb, you have to lift the ENG 1 FIRE pb guard.

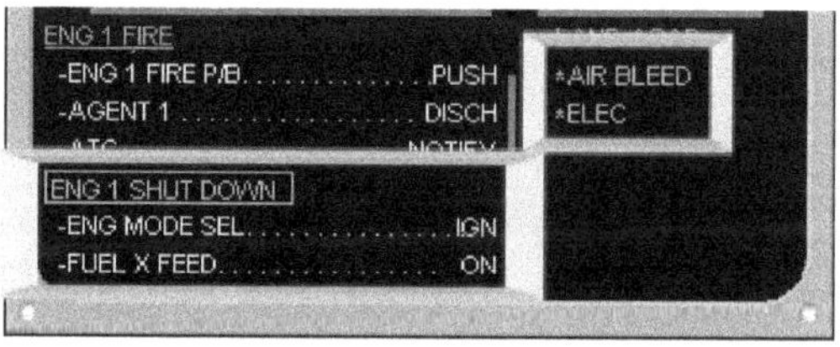

When released, the FIRE pb sw:

* Arms the SQUIBs, shown by the white indications

*Closes FUEL, HYDRAULIC, BLEED and PACK valves of ENG 1.

For this reason, the HYD system is added to the secondary failure list. Simultaneously, an automatic count-down is started on the ECAM. This is to allow the engine to further slow down in order to increase the effectiveness of the agent. The next line of the ECAM procedure calls for the discharge of the fire extinguisher bottle.

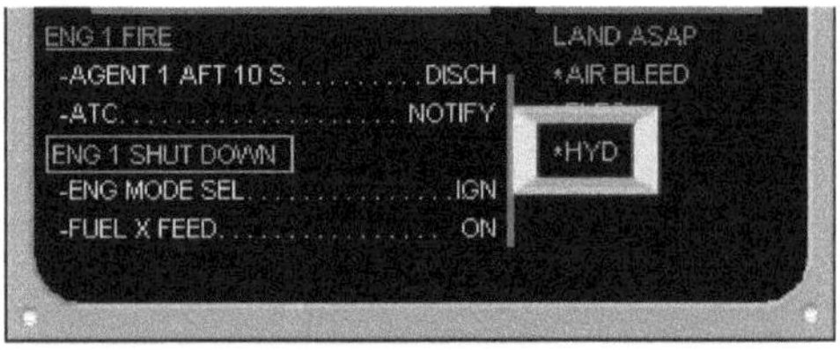

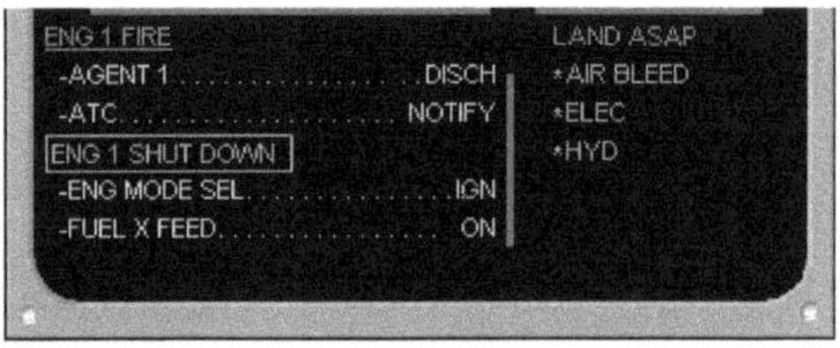

Continue ECAM ACTIONS! The DISCH light illuminates on the AGENT 1 pb. This means the fire extinguisher bottle is depressurized.

It's now time to inform ATC. Since there is no feedback to the FWC, this line will not disappear.

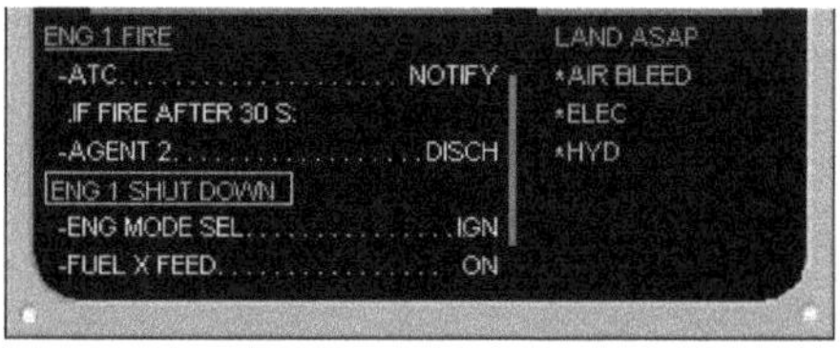

All indications confirm that the fire is still burning. A second count-down now starts. This one is 30 seconds long, in order to give the first agent enough response time. If, at any time, the fire stops, the ENG 1 FIRE message will disappear.

Agent 1 managed to extinguish the fire. The countdown for agent 2 stopped immediately.

Notice: The ENG 1 FIRE pb sw is no longer illuminated. The FIRE light on the ENG panel is now off. The ENG FIRE procedure on the ECAM has disappeared. LAND ASAP has changed from red to amber.

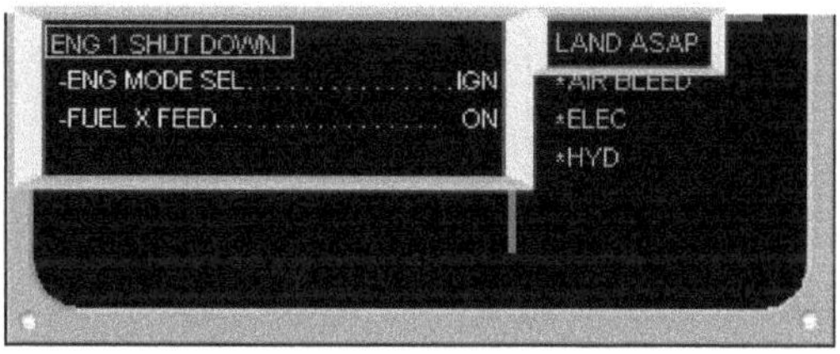

Since the next steps are part of different chapters, we will stop here and proceed with the next failure. You managed a safe landing and changed the aircraft. You are back in cruise, everything is normal, but unfortunately, it is one of those days. suddenly the MW is triggered again.

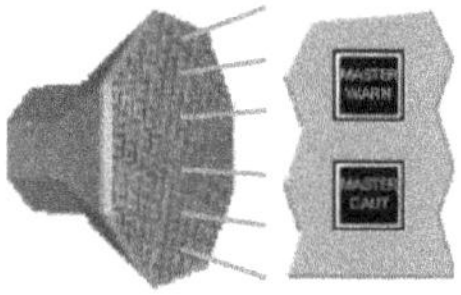

On the E/WD the message FWD CARGO SMOKE with the associated procedure is displayed. Simultaneously a red SMOKE light on the CARGO SMOKE panel illuminates and the COND page is displayed on the SD.

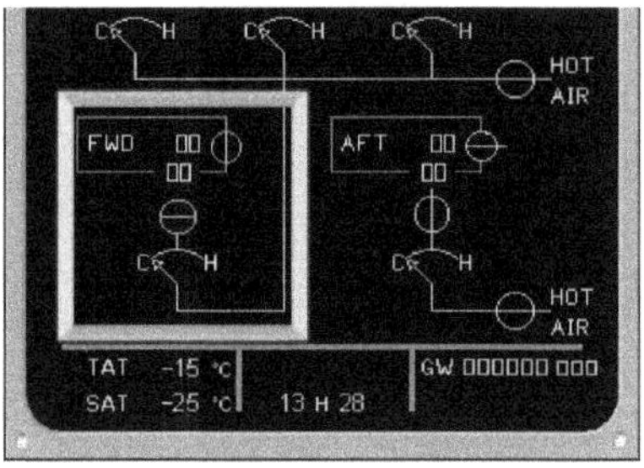

The fwd cargo isolation valves closed automatically. If this had not happened, the corresponding line would have been added to the ECAM procedure. Let's perform ECAM ACTIONS!

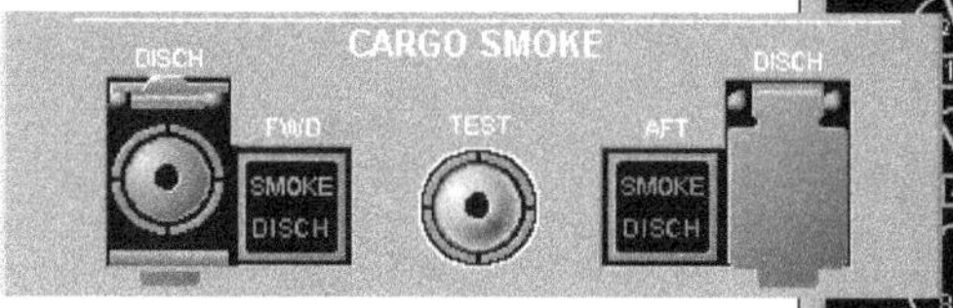

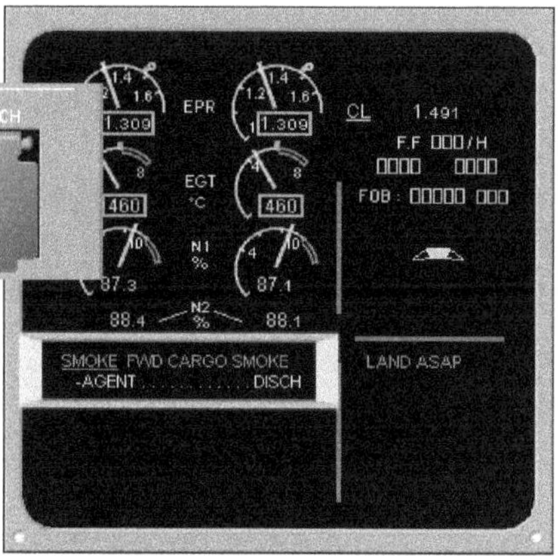

Notice that both DISCH lights illuminate. This is because there is only one fire extinguisher bottle for both cargo compartments. That means there is no fire agent left for a possible AFT CARGO SMOKE warning. Note: Expect the SMOKE warning to remain after agent discharge, even if the smoke source is extinguished. Gases from the smoke source are not evacuated, and smoke detectors are sensitive to the extinguishing agent.

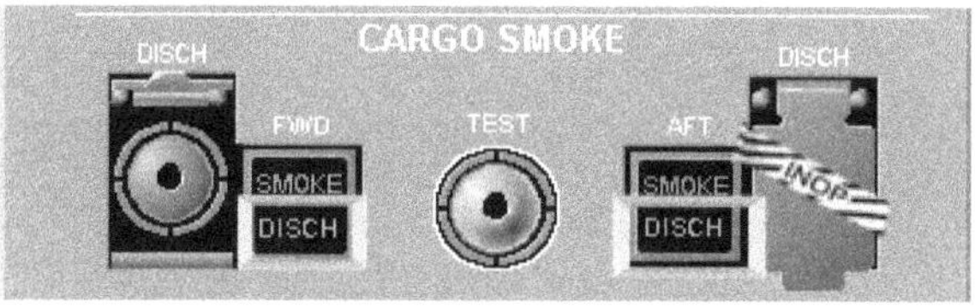

We will stop here and proceed with the next failure. If one of the two fire detection loops of an engine fails, an ECAM message is triggered.

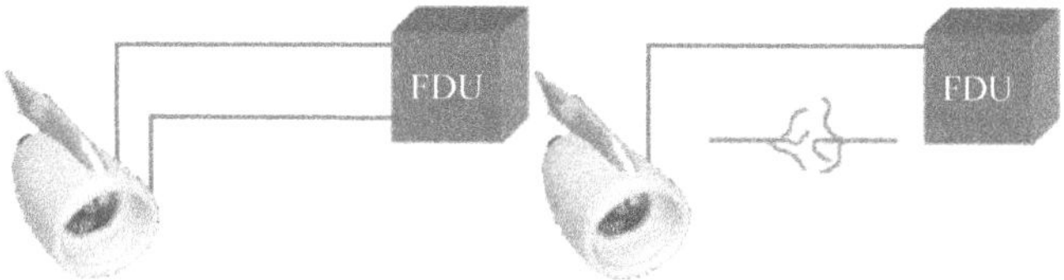

Because of the system redundancy fire detection is still fully operational. Let's see what happens when you lose the second loop. The MC alarm is triggered! Should you lose both loops or the FDU, fire detection is not available anymore for the respective engine or the APU.

Crew awareness and a close monitoring of the respective engine indications are now required. Note: Should both loops break within 5 seconds, a FIRE warning is triggered.

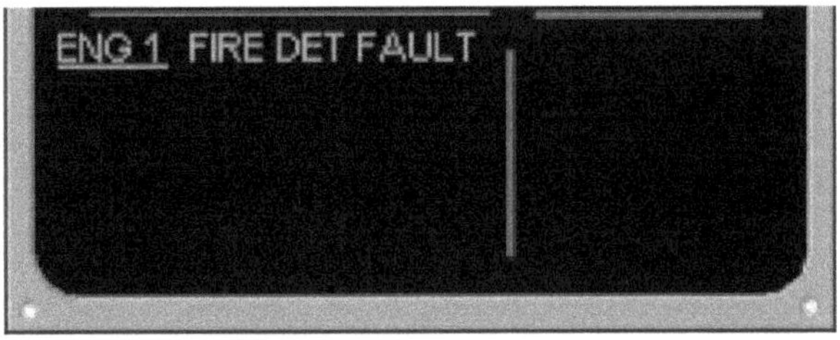

Communication Abnormal Operation

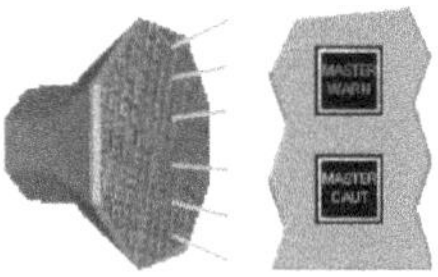

This is a demonstration of continuous VHF transmission. You are the pilot non-flying. You are in cruise, everything is normal. suddenly the MC is triggered!

On the E/WD, read the title of the failure. Observe that there are no ECAM actions and that no system page is displayed for this failure. The VHF 1 radio is continuously transmitting, the most probable cause for this is a stuck PTT switch. Therefore, check all the PTTs.

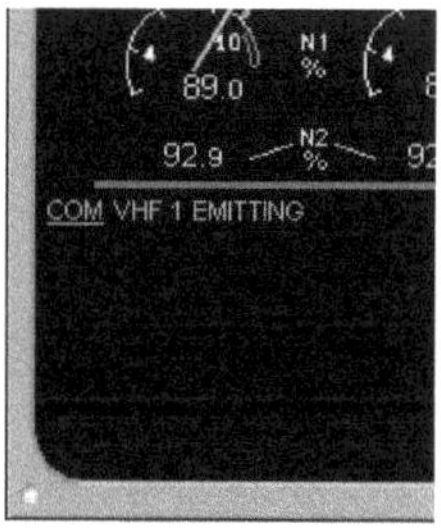

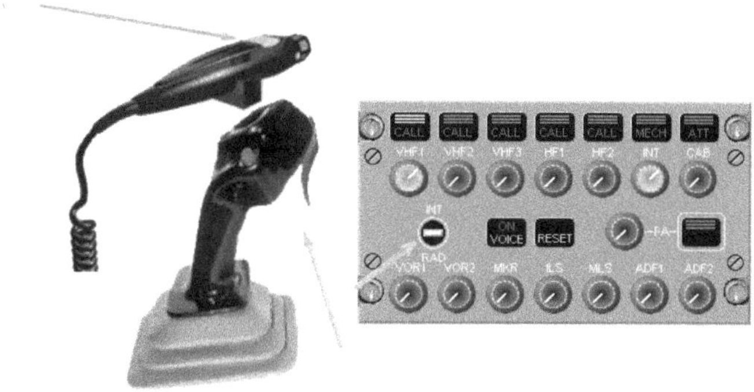

You found one PTT stuck in the “transmit” position. After you release it, the ECAM message disappears. Let’s now look at another communication failure. You are still in cruise and everything is back to normal. ATC asks you to change frequency. When you look at your RMP, however, all the indications are blank. Notice the ON/OFF switch is ON. There is no ECAM caution for this abnormal.

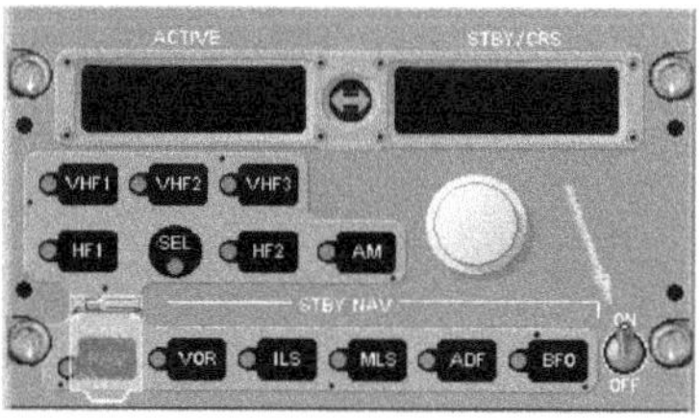

RMP 2 **RMP 3**

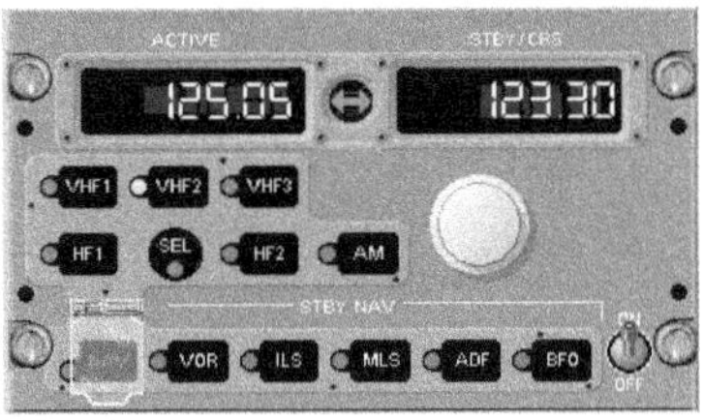

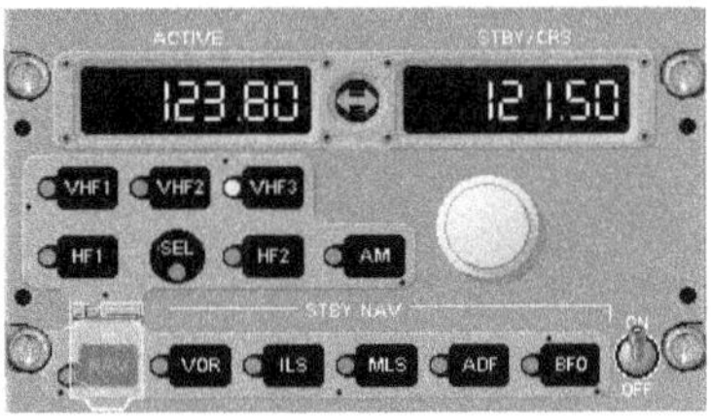

The blank displays with no lights illuminated indicate an RMP failure. All the radio tuning must now be done using RMP 2 or RMP 3. Notice that the SEL light is illuminated because RMP2 is being used to tune VHF 1.

Now let's look at another failure.

You are in the left hand seat. While in cruise, you reach a reporting point on your route. You try to transmit to ATC using your side stick PTT, but you receive no reply. When you check your ACP, you see that all the lights are extinguished. You try to select VHF 1 for transmission, but your selection has no effect.

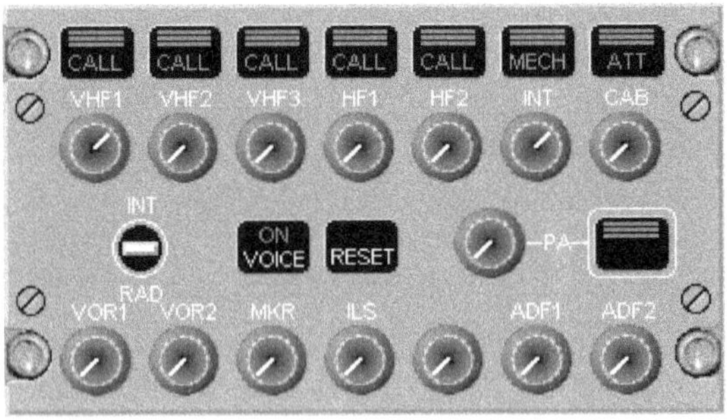

As your were using ACP 1, you can recover audio control through ACP 3 by selecting the CAPT 3 position on the AUDIO SWITCHING selector located on the overhead panel.

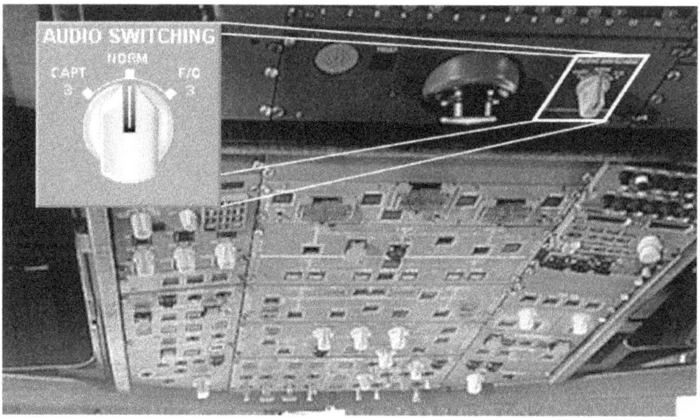

This selection de-energizes ACP 1. The Captain's boomset, oxygen mask and hand-mike transmission and reception functions are now controlled through ACP 3. The AUDIO 3 XFRD message appears on the EWD to indicate that an audio switching selection has been made.

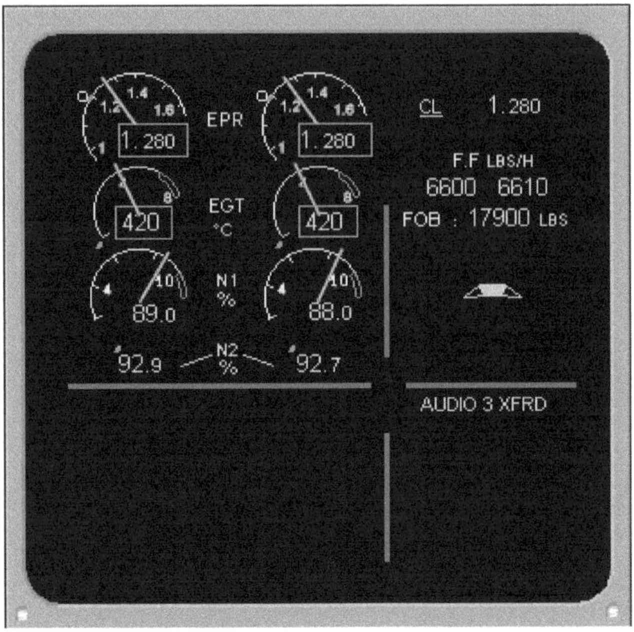
EPR
1.280
1.280
CL 1.280
F.F LBS/H
6600 6610
EGT
°C
420
420
FOB : 17900 LBS
N1
%
89.0
88.0
92.9
N2
%
92.7
AUDIO 3 XFRD

Chapter 4

Abnormal operations IV

Aviation Bookstore
aviation step by step

***NOTE:** information not valid for real flights. Concepts explained as examples for your study and entertainment. For real flights see the original manuals of the manufacturer AIRBUS.*

Aviation Bookstore
aviation step by step

Landing Gear Abnormal Operation

This is a demonstration of a landing gear abnormal operation. You are the PNF. You are in approach, the speed is 200 knots, everything is norma. When you extend the landing gear the MW is triggered!

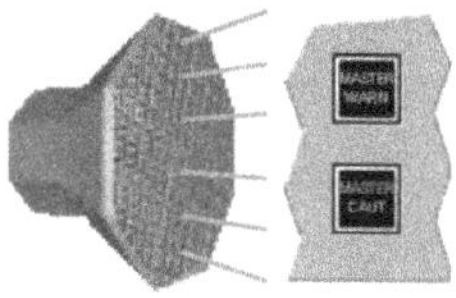

The red warning and associated procedure are displayed on the E/WD. Read the title of the failure.

On the ECAM WHEEL page the landing gear indication is red indicating that the left hand main gear is not down and locked and the message L/G CTL is displayed amber to indicate that the gear and landing gear lever position disagree.

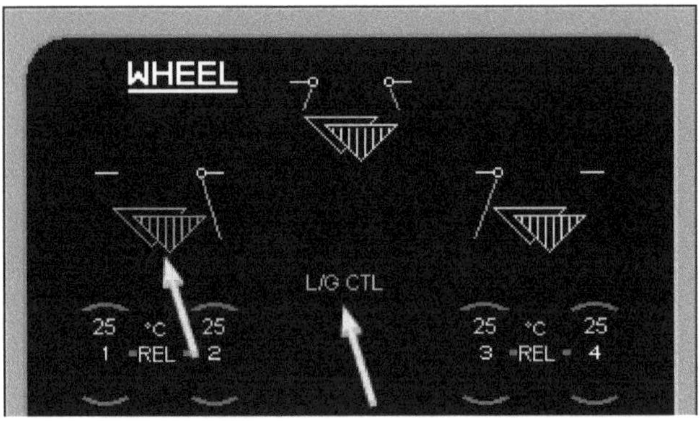

On the landing gear panel the UNLK indication is red indicating that the gear is not locked and the landing gear lever the red arrow is red indicating that the gear is not down and locked.

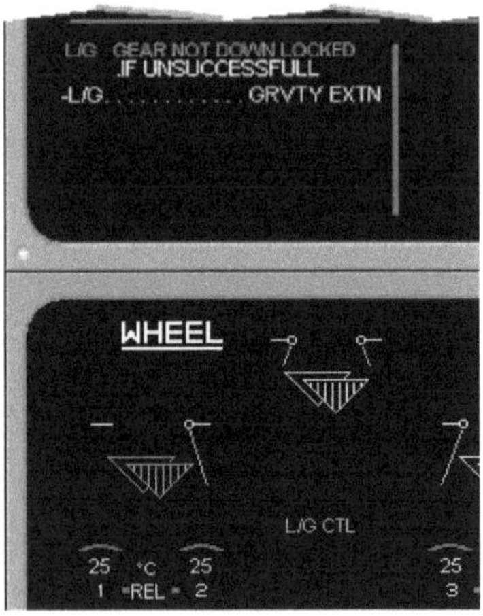

After performing ECAM ACTIONS, the recycle was unsuccessful and the landing gear is still unsafe. The landing gear is extended again and the failure remains. The next ECAM action requires a non ECAM procedure described in the QRH. This non ECAM procedure asks you to extend the landing gear by gravity. To extend the landing gear by gravity the hand crank must be pulled and turned clockwise about 3 turns.

GRAVITY GEAR EXTN handcrank.....PULL and TURN

LANDING GEAR LEVER............................DOWN

GEAR DOWN INDICATIONS.....................CHECK

The handcrank has been turned 3 times. The failure is no longer displayed on the E/WD.Note: after gravity extension the handle should be stowed to allow the PARKING BRK handle to be operated. After gravity extension, the landing gear is down and locked.

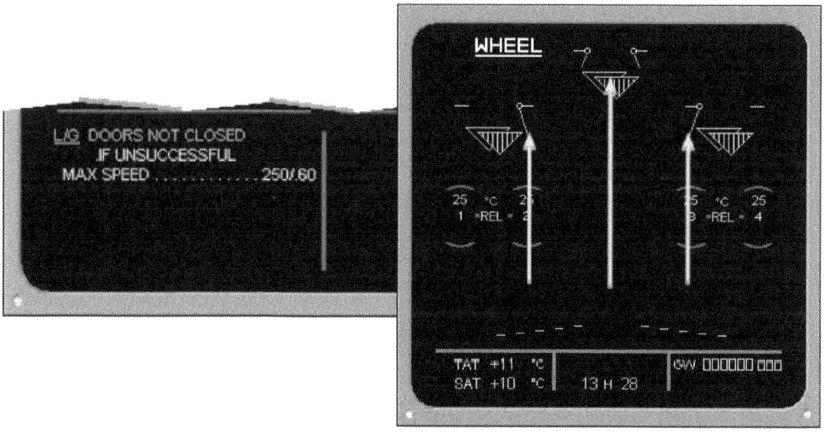

The landing gear doors remain open. On the ECAM WHEEL page, this is indicated by the amber door indications in open position and by the failure message on the E/WD. The Status page is displayed. Approach CAT 3 SINGLE only is listed in the left column. CAT 3 DUAL and NW STEER are inoperative as displayed in the right column.

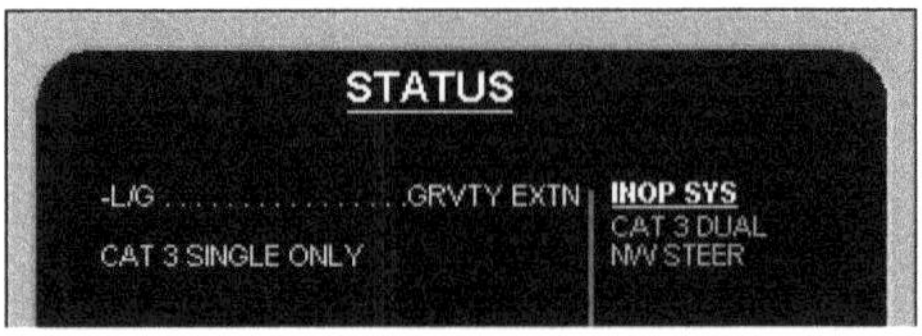

In the same way, other failures can be displayed on the ECAM WHEEL page. Let's see.

In this example one system (LGCIU 1) detects the nose landing gear in transit and displays the triangle in red. The other system (LGCIU 2) detects the nose gear down and locked and shows a green triangle. On each gear, only one green triangle is sufficient to indicate that the gear is down and locked.

On the main gear, one triangle is missing meaning that the gear is detected up and locked by one system (LGCIU 2). The other triangle is red because the other system (LGCIU 1) detects the gear in transit.

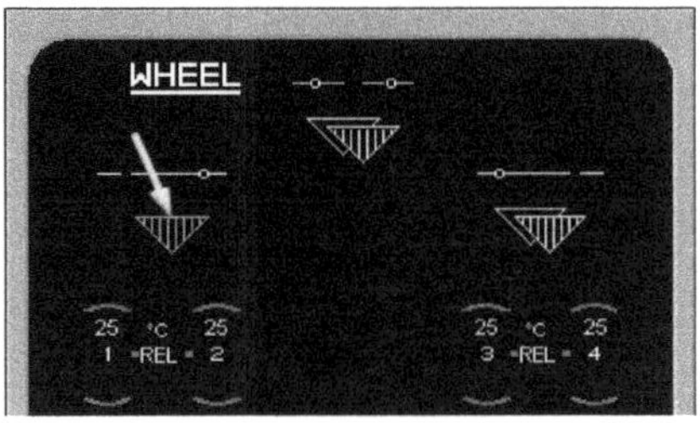

The message UP LOCK appears when the landing gear is down and locked with a gear up lock engaged. It is associated with an ECAM caution.

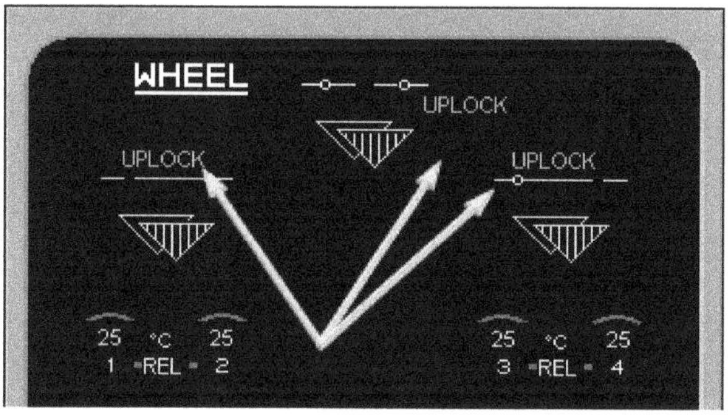

The message STEERING appears in case of nose wheel steering failure. It is associated with an ECAM caution.

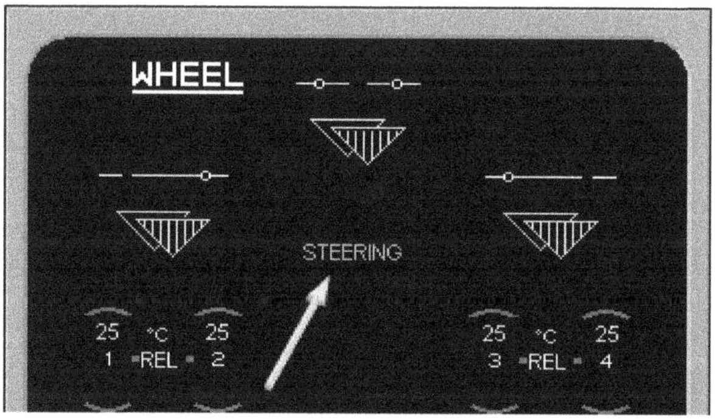

Now, let's see another failures. You are the PNF. You are just about to touch down, everything is normal…

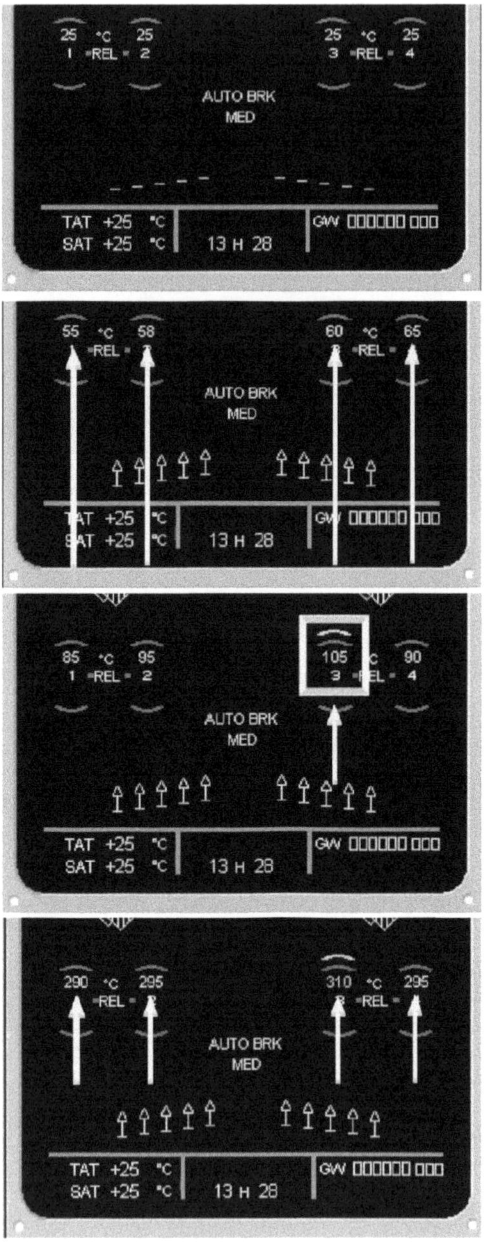
25 °C 25
1 REL 2
25 °C 25
3 REL 4
AUTO BRK
MED
TAT +25 °C
SAT +25 °C
13 H 28
GW
55 °C 58
REL
60 °C 65
REL
AUTO BRK
MED
TAT +25 °C
SAT +25 °C
13 H 28
GW
85 °C 95
1 REL 2
105 °C 90
3 REL 4
AUTO BRK
MED
TAT +25 °C
SAT +25 °C
13 H 28
GW
290 °C 295
REL
310 °C 295
REL
AUTO BRK
MED
TAT +25 °C
SAT +25 °C
13 H 28
GW

EIS ECAM Abnormal Operation

The two identical Flight Warning Computers (FWC) receive data from: the aircraft system sensors to generate red Warnings, and the SDAC to generate amber Cautions. The FWC then supply; the DMC for the display of alert messages, the attention getters, and the loudspeakers for aural alerts and synthetic voice messages.

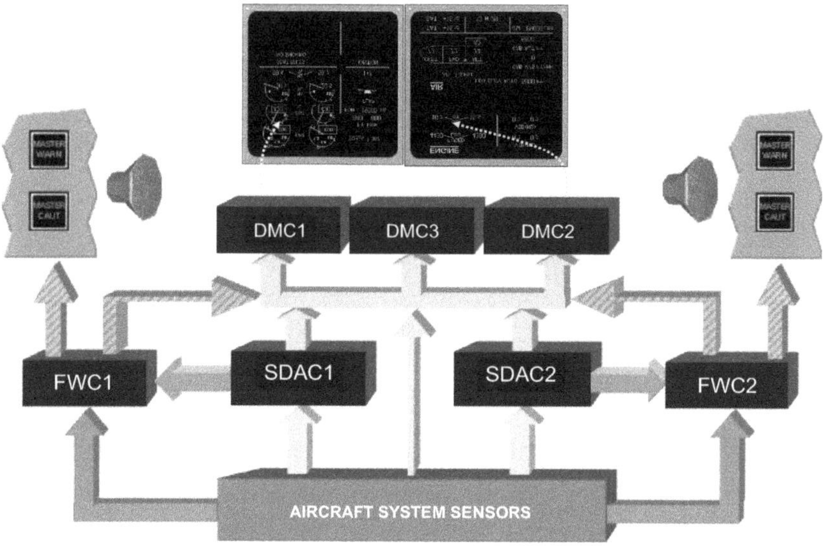

To begin we will look at what happens when one or both Flight Warning Computers fail. For a single FWC fault a caution message will be displayed on the E/WD. Notice that there are no blue action lines, so no crew action is required. No system page is called because there is no system page for EIS.

On the STATUS page you will notice that the aircraft's autoflight capability is downgraded to CAT 3 SINGLE. The inoperative systems are CAT 3 and FWC 1. After a review of the STATUS page by both pilots the ECAM procedure can be completed.

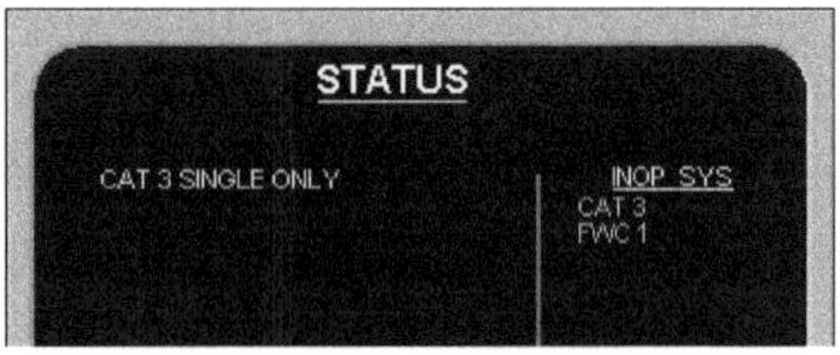

The CRUISE page is displayed on the lower ECAM screen and the ECAM control panel has no lights on it. A boxed STS reminder appears on the E/WD to remind you that there is a message on the STATUS page. That has completed the procedure for a FWC 1 fault.

As a result of the failure if any other WARNING or CAUTION is generated, that would light an indicator, you will notice that only half of lights are illuminated. In this example with a FWC 1 fault only the lower part of the Captains' indicators and the upper part of the First Officers' indicators are illuminated. Note; if FWC2 had failed the lights would be reversed.

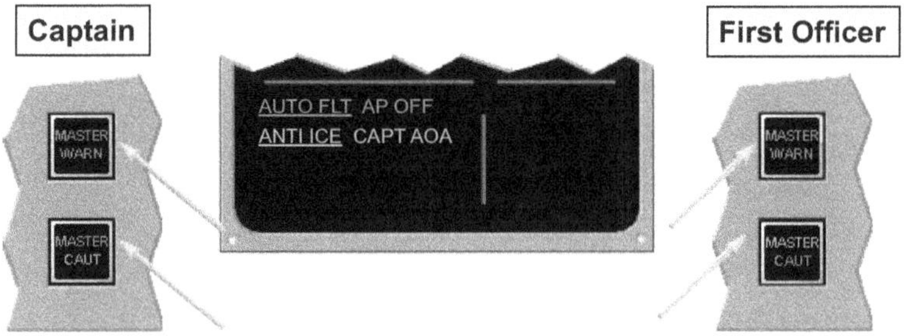

The loss of both FWC will remove the automatic monitoring of the aircraft systems. A message and information, generated by a DMC, is displayed at the bottom of the E/WD. There will be no further ECAM Cautions or Warnings, aural warnings, MASTER CAUTION or MASTER WARNING lights.

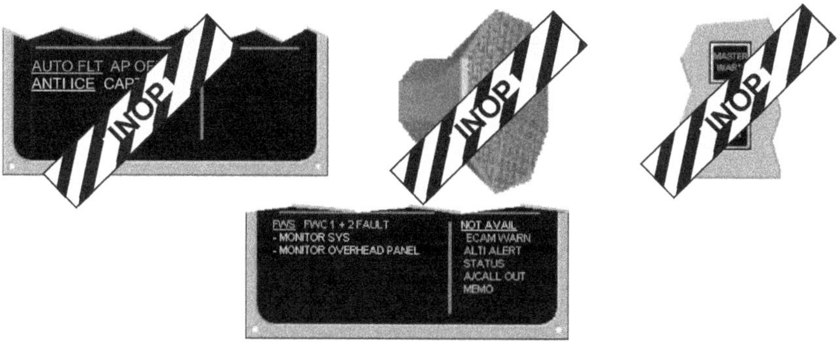

The action lines advise the pilots to monitor the ECAM system pages and the overhead panel for local warnings. Notice that there is a NOT AVAIL list on the right hand side. The NOT AVAIL items are: ECAM Warnings, Altitude Alerts, the STATUS page, Automatic call outs and Memos.

The overhead panel should be checked for any local warnings. The system pages can be checked for abnormal indications either by individual selection of the pages or by using the ALL pb to sequence through the pages. We will hold down the ALL pb for you to show you what happens. Note: if the ALL pb is released, part way through the sequence, the CRUISE page is displayed.

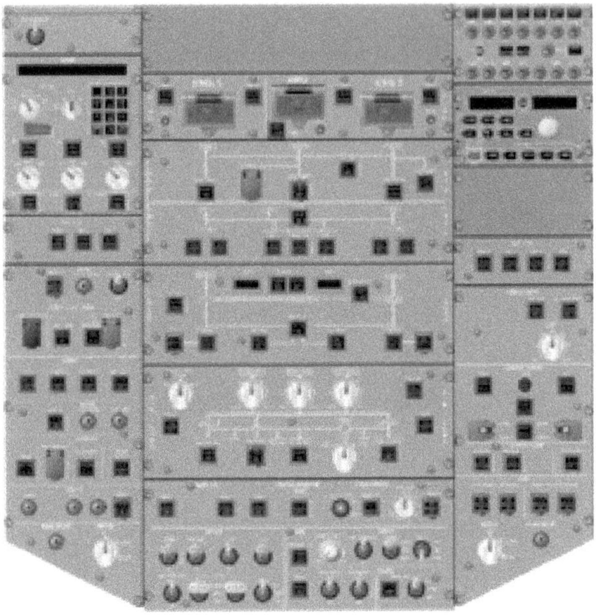

To finish this module we will briefly look at the use of the Emergency Cancel pb. It is normally used to cancel an intermittent nuisance message. It may be used, if necessary, to cancel most aural warnings. Let's look at an example of an intermittent nuisance message.

When an intermittent caution occurs the normal aural and visual alerts will be triggered. The EMER CANC pb can be used to tell ECAM to suppress the caution. As a result of pressing the EMER

CANC pb a white message appears on the E/WD advising you that the caution has been cancelled.

Ventilation Abnormal Operation

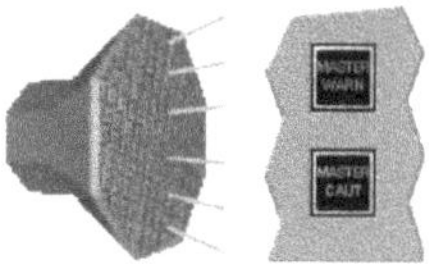

This is a demonstration of an EXTRACT valve fault. You are in flight and everything is normal, suddenly the MW **ENG 1 FIRE** is triggered!

Let's look at the fault indications;

* A fault message and associated checklist on the E/WD.

* The ECAM CAB PRESS page has been automatically called to show amber indications

* A FAULT light has illuminated on the VENTILATION panel

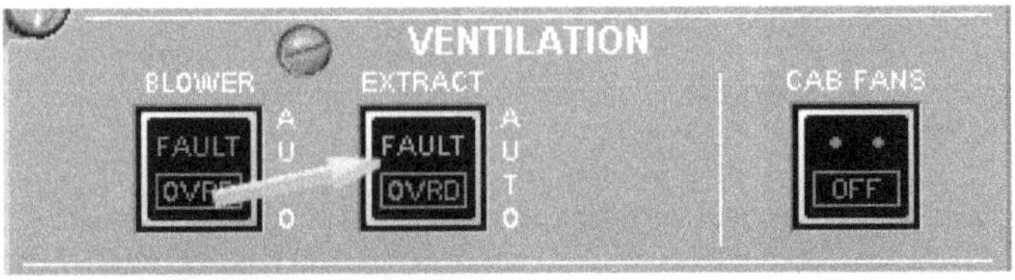

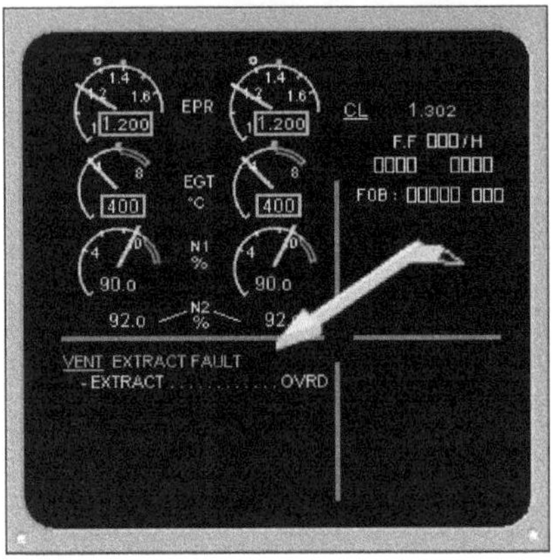

On the E/WD, read the title of the failure. Before commencing the ECAM actions, let's look at the CAB PRESS page indications first. In this example VENT and EXTRACT messages are shown in amber telling you that there is a problem with the extract system. You, as PNF Read and Perform the ECAM Actions. When you switch to OVERRIDE;

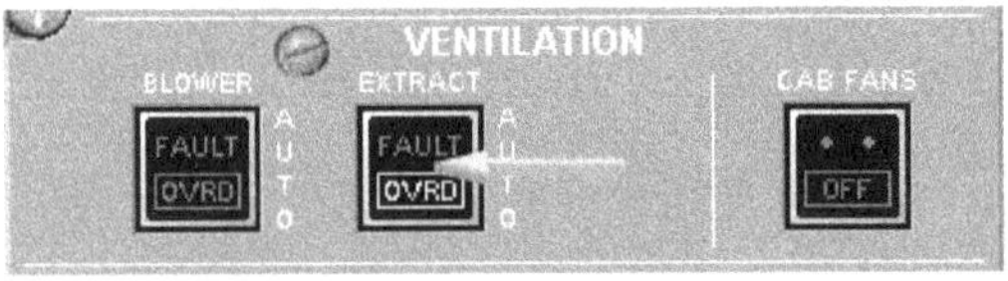

* An OVRD light illuminates in the switch,
* The FAULT light extinguishes,
* The blue action line on ECAM disappears.

There are no changes to the indications on the system display but by pushing the EXTRACT switch to OVERRIDE the ventilation system has been reconfigured to ensure adequate ventilation of the avionics system.

The next step in this example, after checking with the other pilot, is to clear ECAM. The Status page is now displayed and the E/WD returns to a normal display.

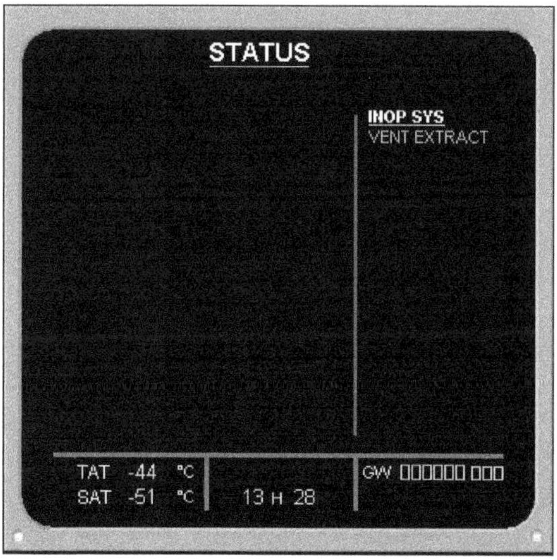

The inoperative system is ventilation extraction. The STATUS page is replaced with the CRUISE page. Notice that there is a STS reminder at the bottom of the E/WD to remind you that there is something in the STATUS page. This completes the ECAM procedure for a VENT EXTRACT FAULT

Let's continue with another abnormal indications associated with the Ventilation system. The FAULT lights on either the BLOWER or the EXTRACT switches illuminate for: low pressure, computer problems or as a smoke warning.

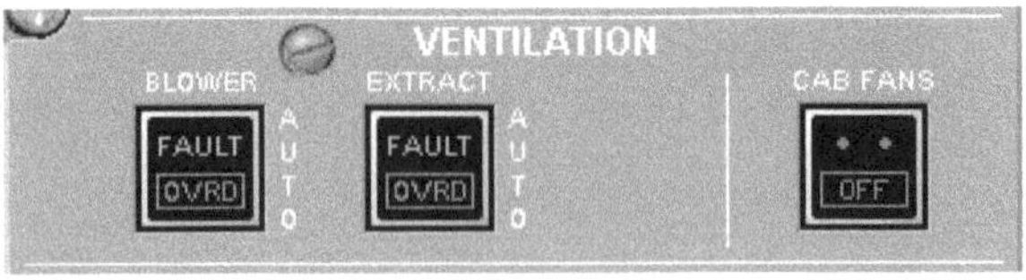

In addition, the BLOWER FAULT light illuminates as a warning of a duct overheat.

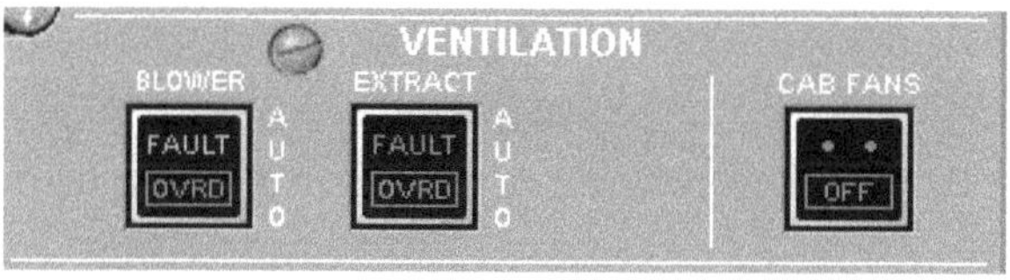

These abnormal cases will also be reflected on the ECAM CAB PRESS page depending on the circumstances. In the example shown both the INLET and EXTRACT indications have changed to amber. Since this is a fault with the ventilation system the VENT indication also changes to amber.

The OVERRIDE mode is used during ECAM abnormal procedures. When the BLOWER sw is in OVERRIDE mode, the

BLOWER fan is turned off. The system goes into closed configuration with air from the Air Conditioning system added to the ventilation air.

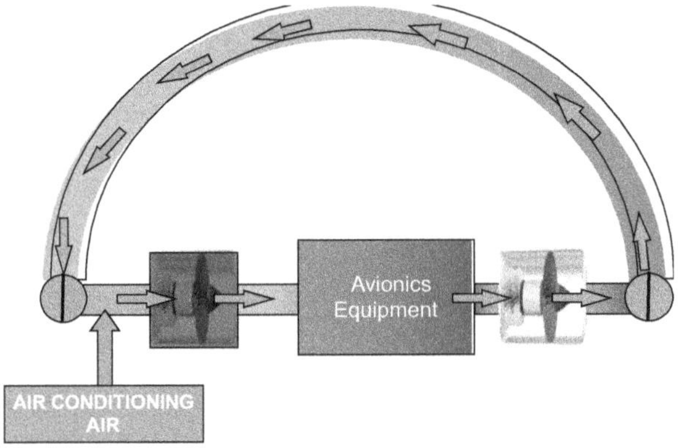

When the EXTRACT sw is in the OVERRIDE Mode, a reconfiguration of the ventilation system occurs to ensure adequate avionics equipment cooling.

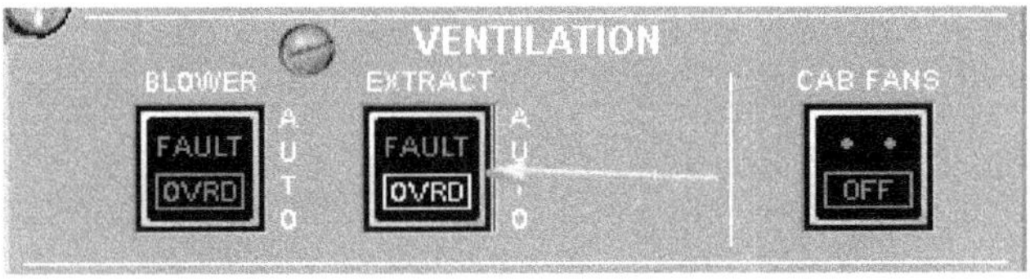

There is an ECAM procedure that requires both switches to be set to OVRD. This further reconfigures the system, but on this occasion, there is a change on the CAB PRESS system display.

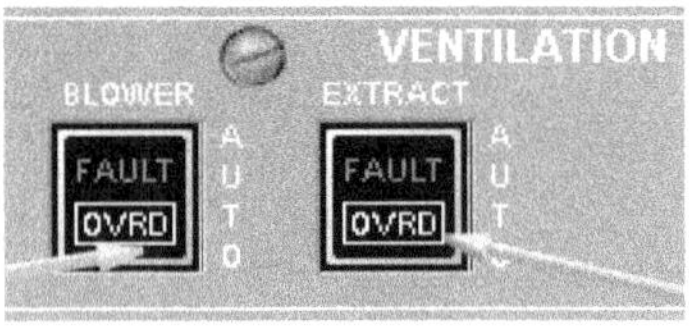

The EXTRACT valve indication shows partly open. This is telling you that a small internal flap has opened allowing air from the ventilation system to be discharged overboard. This configuration is used in the AVIONICS SMOKE procedure.

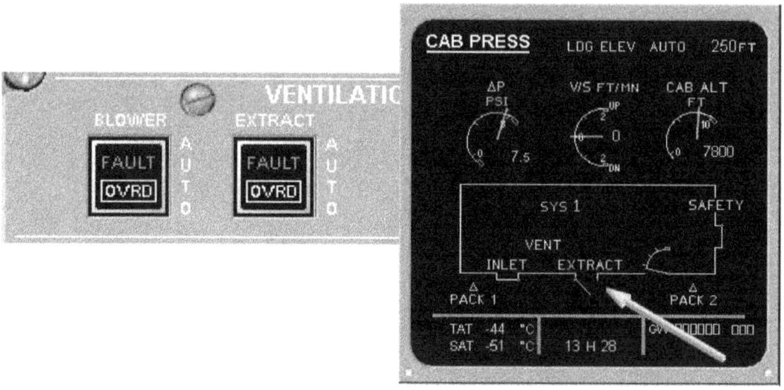

Pressurization Abnormal Operation

This is an example of an abnormal procedure. We will use a CABIN PRESSURE FAULT where one of the two automatic systems fails. At this time, the aircraft is in cruise and all systems are functioning normally. When the ECAM system detects a failure, the appropriate caution is generated and the ECAM CAB PRESS system page is automatically called. Due to the fact that this is only a crew awareness, there was no Master Caution.

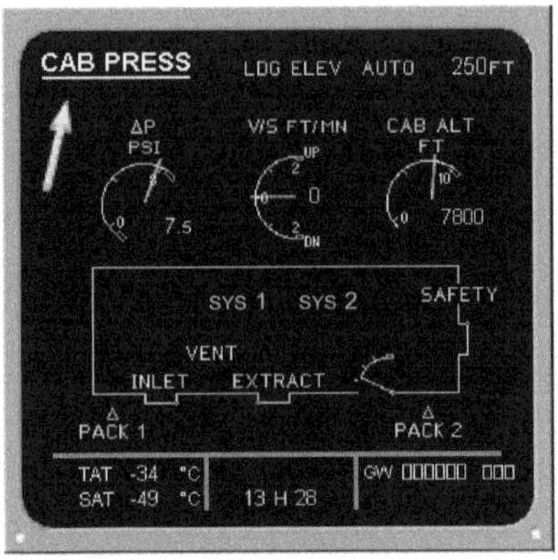

The failure message is read. In the case of a CABIN PRESSURE SYSTEM 1 FAULT, there is no action required by the crew. System two is automatically activated and controls the outflow valve. Once both pilots have reviewed the Status page, clear the Status Page.

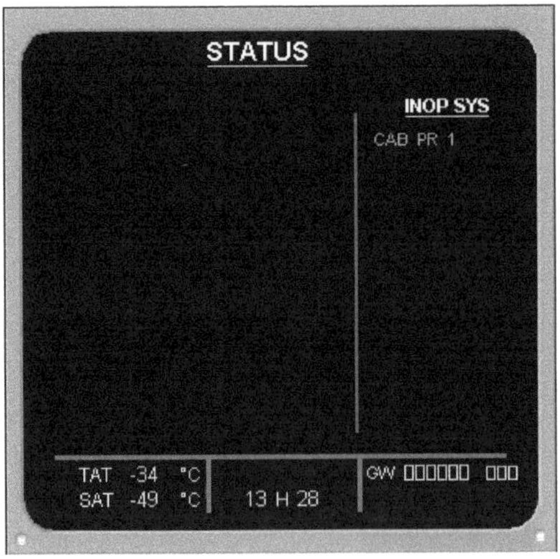

Let's look at a second failure to illustrate the functionality of the system, the associated indication, and the ECAM procedure. We will use a Dual System Fault to illustrate these procedures. The aircraft is in cruise at FL350. Suddenly the MC alarm is triggered!

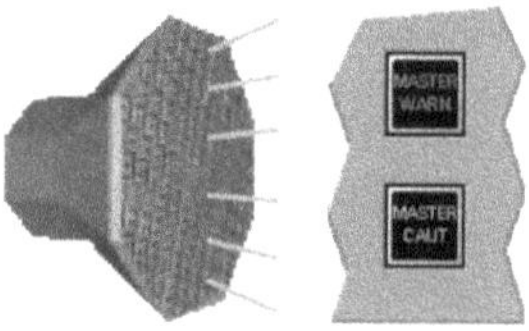

The ECAM system has generated a failure message, with action lines, and called the ECAM CAB PRESS page. Read the ECAM failure title and on the ECAM CAB PRESS page observe the indications shown in amber.

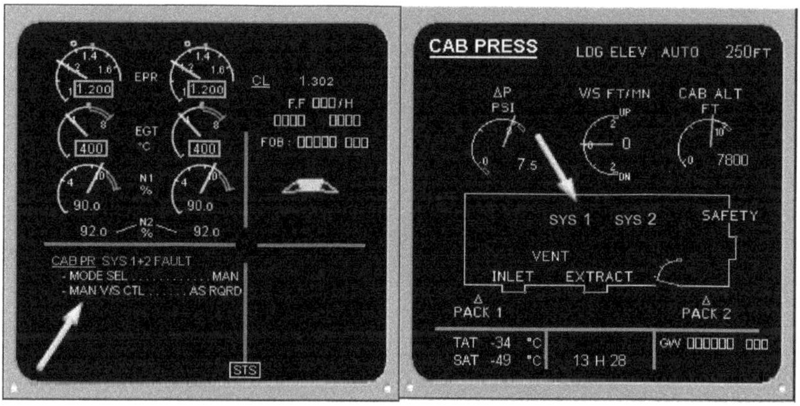

Read and complete the action items on the E/WD. In this case, both automatic systems have failed, we must switch to manual

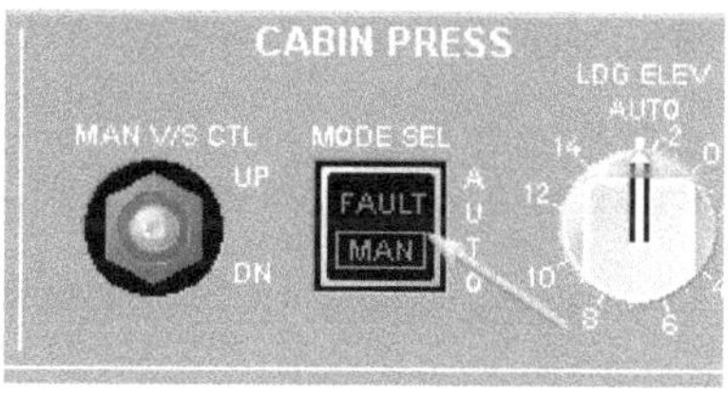

mode. Notice that the MODE SEL switch has an amber FAULT indication to help you locate it.

The indication on the switch changes to show MAN in white and a green MAN message appears on the system page to indicate that the pressurization system is in manual mode.

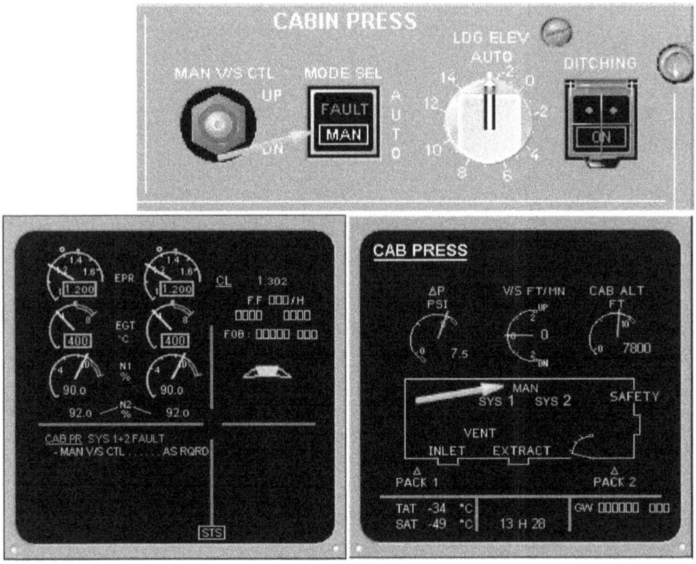

You must now control the pressurization of the aircraft by using the MAN V/S CTL switch to move the outflow valve. Notice also that, since you have control of the outflow You must now control the pressurization of the aircraft by using the MAN V/S CTL switch to move the outflow valve. Notice also that, since you have control of the outflow valve, the Landing Elevation indication is no longer displayed. Let's pause the ECAM procedure and look at the use of the MAN V/S CTL. The MAN Vertical Speed switch is spring loaded to neutral. If you hold the switch in the up position the Outflow Valve slowly opens, the Vertical Speed will change, and the Cabin Altitude will increase.

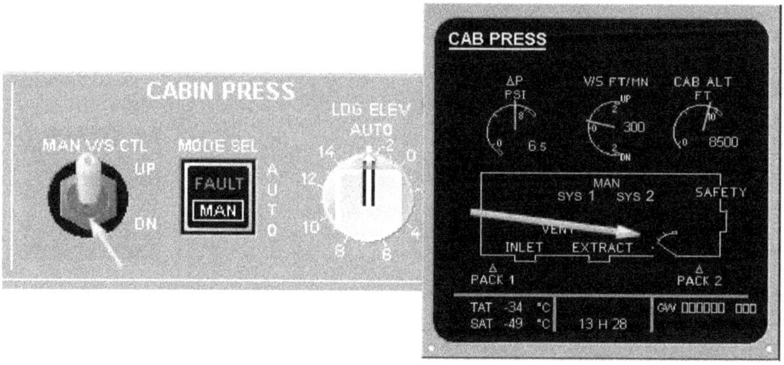

When the MAN V/S CTL is returned to the neutral position the outflow valve will stop moving. The Vertical speed will stabilize and the cabin altitude will change. Conversely if you hold the switch down the Outflow Valve will close, the Vertical Speed will change and the Cabin Altitude will decrease. Now let's return to the ECAM procedure.

Notice that the Vertical Speed indication is zero. This means that there is no need to adjust the outflow valve position using the MAN V/S CTL. Because you have to continually use the MAN V/S CTL switch the action line is not removed.

The STATUS page provides the pilots with information; target Vertical Speeds, target cabin altitudes, which depend on the flight level and a final approach procedure to set the V/S CTL fully up. This ensures that the outflow valve opens completely and the aircraft depressurizes. For our example we are at FL350 so the target cabin altitude is 6500 feet. Review the inoperative systems in the right hand column.

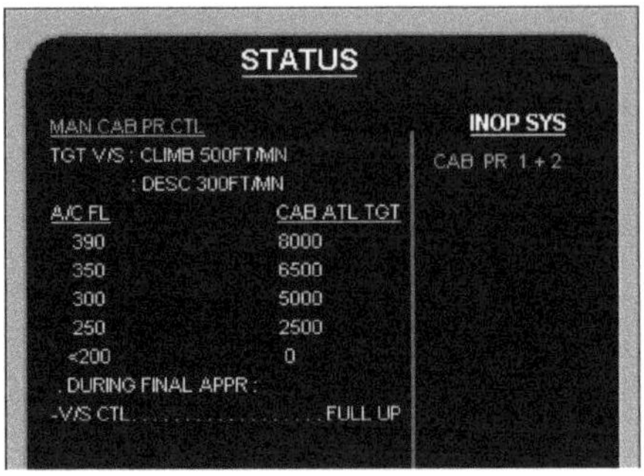

Remember that information will remain on the Status Page for the duration of the flight and can be reviewed at any time. Notice that on the CRUISE page, the indication of cabin vertical speed has changed to a gauge format. This along with the cabin altitude indication can be used to set the correct position of the outflow valve. The ECAM procedure for a Cabin Pressure System 1+2 Fault is complete.

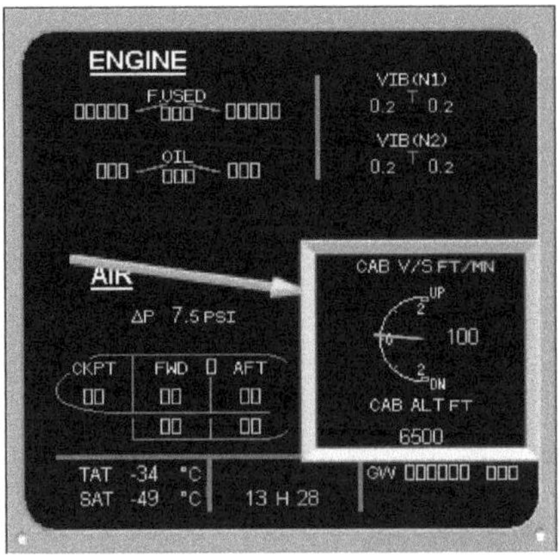

Flight Controls Abnormal Operation

Let's first see the redundancy aspects of the flight controls. In case of a computer failure, the second computer of the same type takes over (ELAC, SEC and FAC cases).

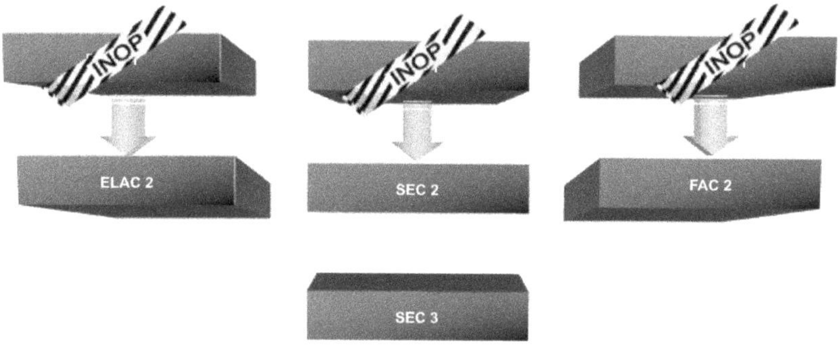

Note that SEC 3 is only used for spoiler control and does not participate (as SEC 1 and 2) in the flight control system reconfiguration. Each primary flight control surface is supplied by 2 (ailerons, elevators and stabilizer) or 3 (rudder) different hydraulic sources. Each control surface has an independent actuator for each hydraulic source. Each actuator is controlled by a different computer. One actuator for each surface is always in active mode. The other stays in damping mode and is monitored by its associated computer. We will demonstrate these points in the following ELAC 1 failure. Assume, you are the PNF. You are in flight and everything is normal. Suddenly the MC alarm.

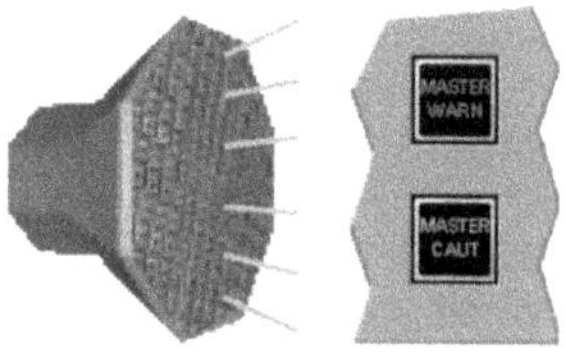

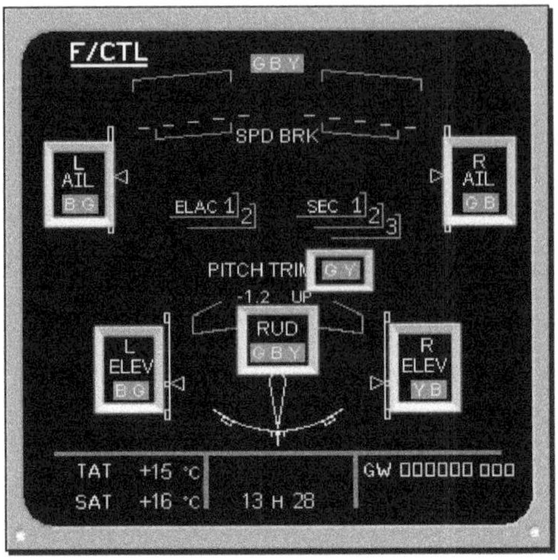

On the E/WD, read the title of the failure.

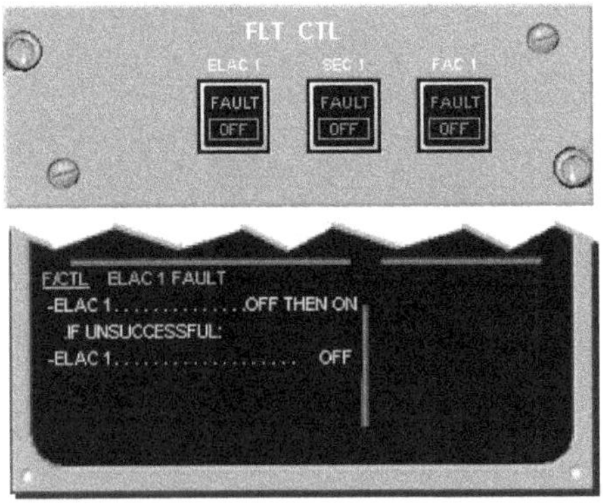

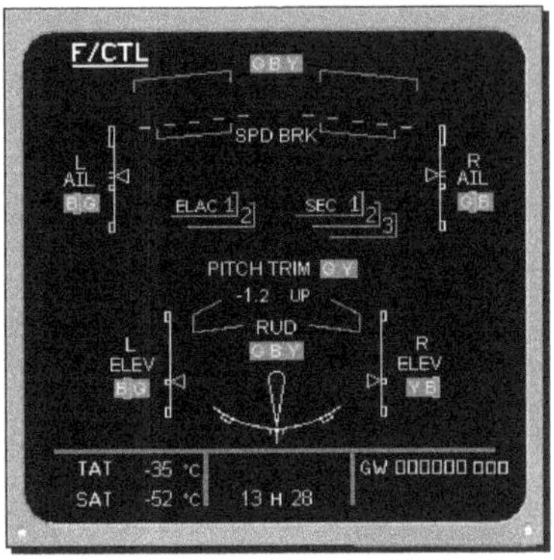

Observe ELAC indications on the ECAM F/CTL page. The "1" and the box of ELAC 1 are in amber. The "2" of ELAC 2 remains in green indicating that ELAC 2 has taken over automatically. Indications for aileron and elevator actuators, controlled by ELAC 1, are partially boxed in amber. This means that these actuators have automatically switched to damping mode.

Actuators controlled by ELAC 2 have automatically switched from damping to active mode to control the ailerons and elevators. They are not boxed on the ECAM F/CTL page. ELAC 1 FAULT light is illuminated on the FLT CTL panel. We can see that the first action is to reset the faulty computer. The PF will ask you to perform the ECAM actions.

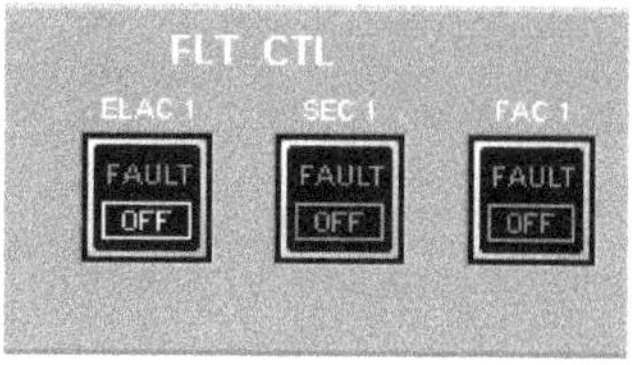

ELAC 1 has not been successfully reset. The procedure requires you to switch ELAC 1 OFF.

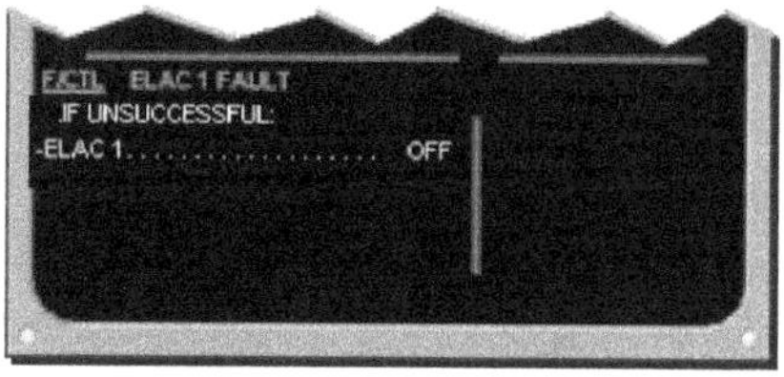

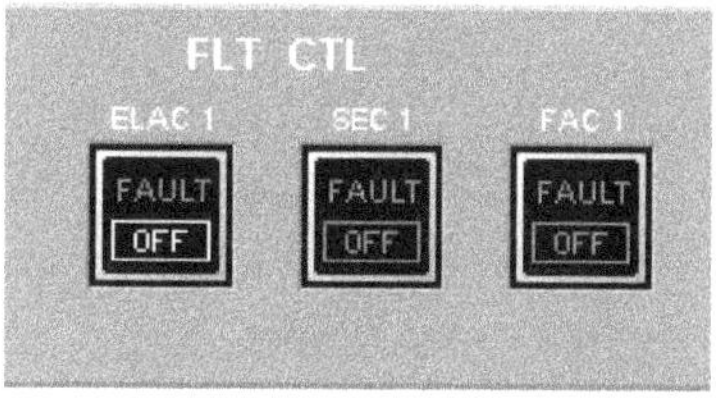

The STATUS page is displayed for review. A green message advises us that we have a degradation in approach capability. Note that there are

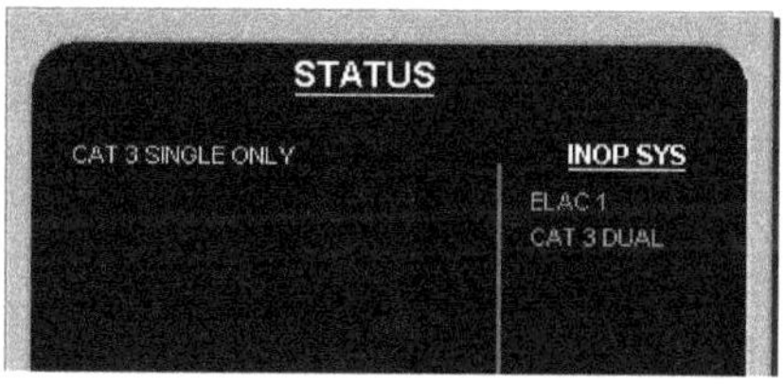

loss of neither control surfaces nor maneuvering ability. You can see, in the INOP SYS column, that ELAC 1 and CAT 3 DUAL are now inoperative.

Now, we will now review the other redundancy features and protections we did not have the opportunity to see when performing the previous procedure. In case of complete failure of ELAC computers, SEC 2 computer takes over the control of the elevators and the THS. The roll function is still available via the spoilers. The ailerons are in damping mode.

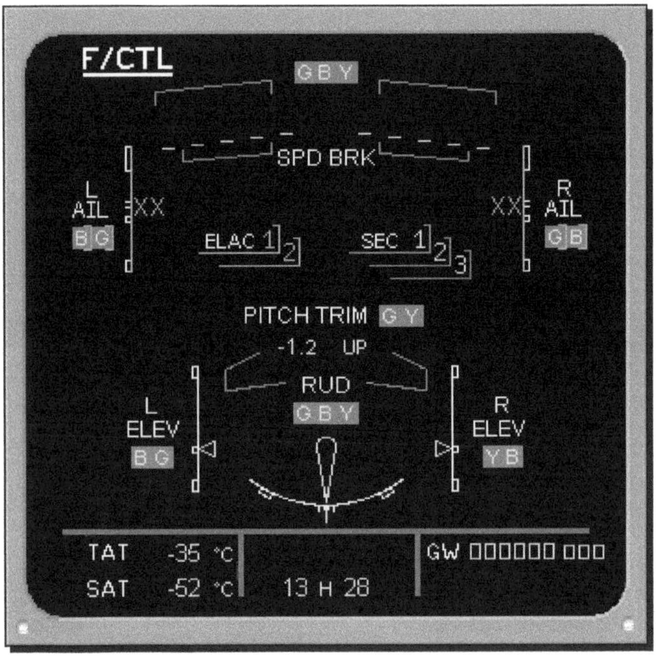

In the very low probability of SEC 2 failure, it will be replaced by SEC 1.

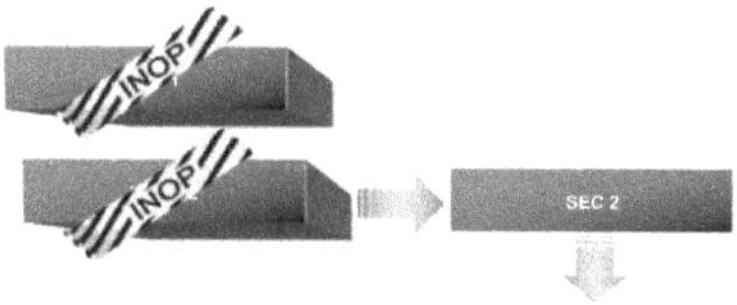

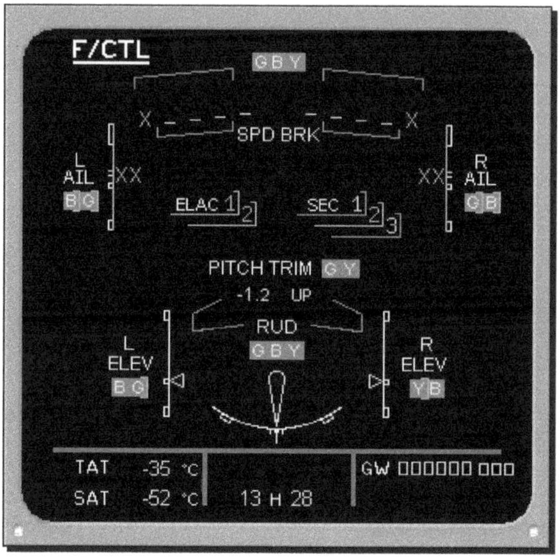

Another case of failure is an elevator fault. In this case, the deflection of the remaining one is limited to avoid excessive asymmetrical loads on the tail or rear fuselage and to reduce the asymmetry

effects. The remaining surface is sufficient to resume normal flight. In case of total loss of normal control or hydraulic supply, the elevators are automatically set to and maintained in the neutral position.

Provided the appropriate Hydraulic systems are available:

- Mechanical control of the THS is always available via the pitch trim wheels.

- For the rudder, mechanical control via the rudder pedals is always available even in case of complete failure of FAC computers.

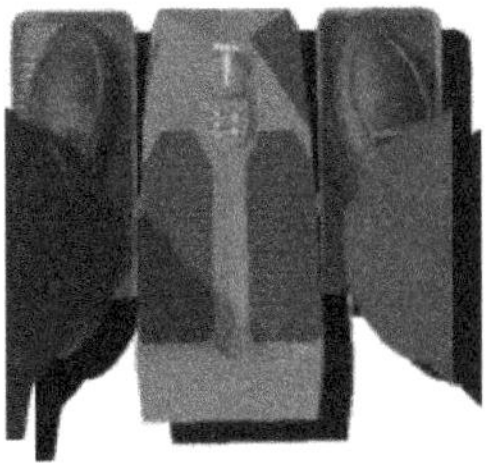

This is called the mechanical backup.

General rules: If a computer fails, the second computer of the same type takes over. Each primary flight control surface is supplied by at least 2 different hydraulic sources. An independent actuator is associated with each hydraulic source for each surface. Each actuator is controlled by a different computer. For a surface, one actuator is always in active mode while the other(s) are in damping mode.

Ailerons/Elevators/THS/Rudder: If both ELAC fail, SEC 2 controls the elevators and the THS. The ailerons are in damping mode. If SEC 2 fails, it will be replaced by SEC 1. In case of one elevator failure, the deflection of the remaining one is limited to avoid excessive loads on the tail and rear fuselage. Mechanical backup is always available using the rudder pedals and pitch trim wheels, provided the appropriate hydraulic systems are available. Let's see how the spoilers are controlled by the SEC. The spoiler control is shared by the 3 SEC.

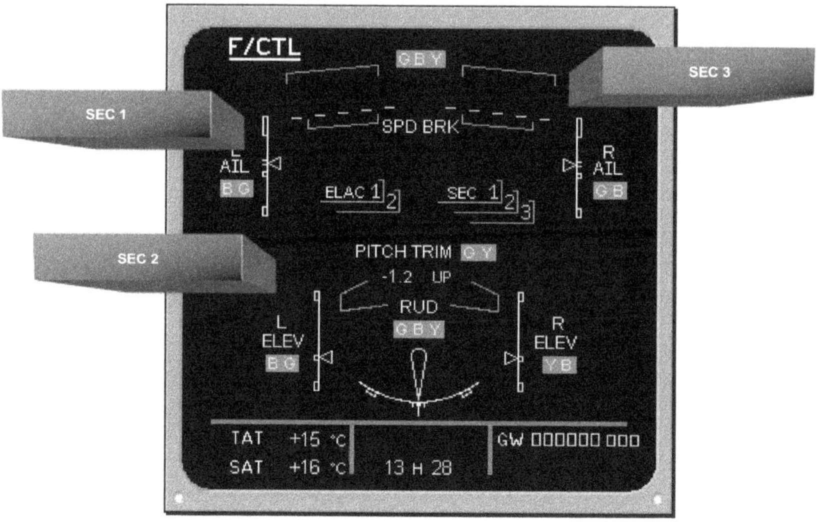

We will now see a spoiler fault. Assume we are in descent, speed brake extended and no previous fault has occurred. Suddenly the MC again

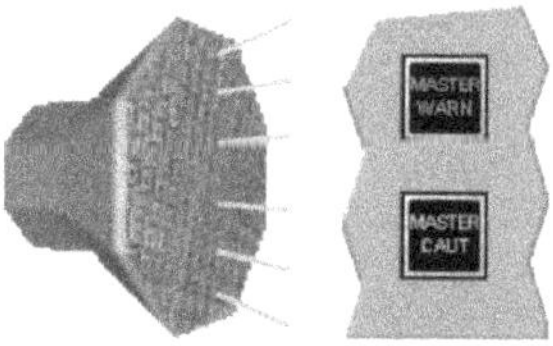

On the E/WD, read the title of the failure. Observe left spoiler 3 indication in amber on the ECAM F/CTL page.

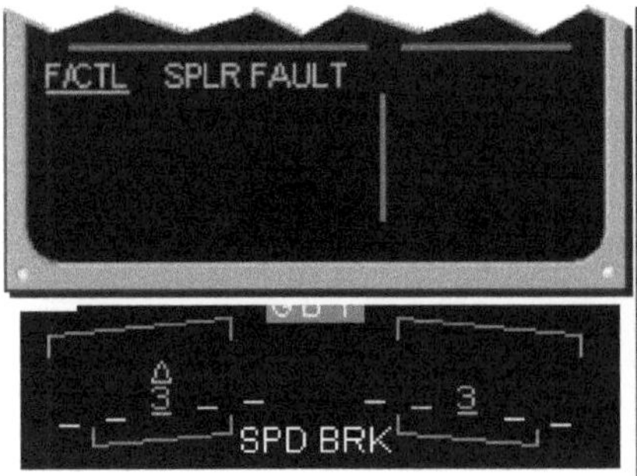

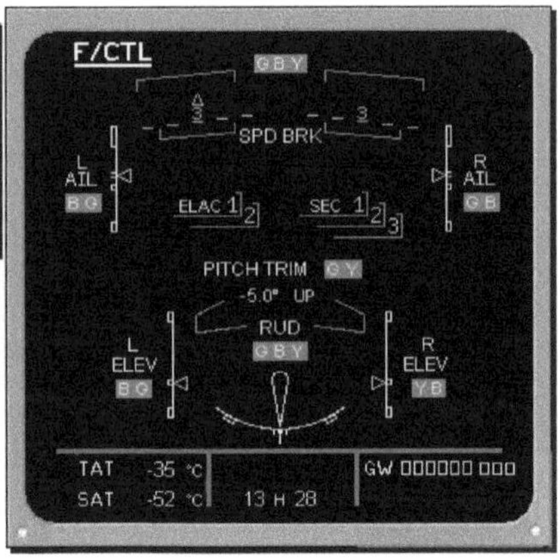

These indications mean that left spoiler 3 has failed in the extended position. The SEC, which controls this spoiler and has detected the fault, will automatically retract it and inhibit its symmetric one on the other wing, right spoiler 3, also shown in amber. You can now observe on the ECAM F/CTL page that spoilers 3 on both wings are retracted and shown in amber. Note that when a spoiler has failed on one wing, the symmetric one is automatically inhibited. As a general rule, spoilers are automatically retracted when faulty or not electrically controlled. As no action is required in the procedure, after review and confirmation from the PF, check status.

The STATUS page is displayed for review. A blue message advises us that we will have to calculate a new landing distance. This is due to the fact that both spoilers 3 will stay retracted when the remaining ground spoilers extend at touch down.

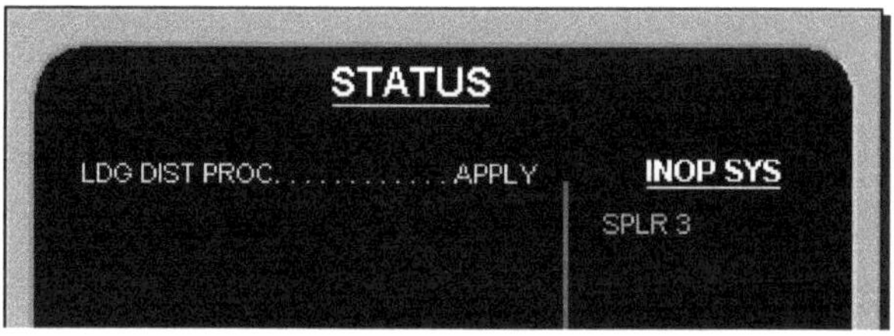

We will now review the other spoiler aspects we did not have the opportunity to see when performing the previous procedure. In certain circumstances, speed brake extension can be inhibited. In these cases, the spoilers are automatically retracted and stay in this position until the inhibition conditions no longer exist and the lever is reset. A Master caution and an ECAM procedure are also triggered.

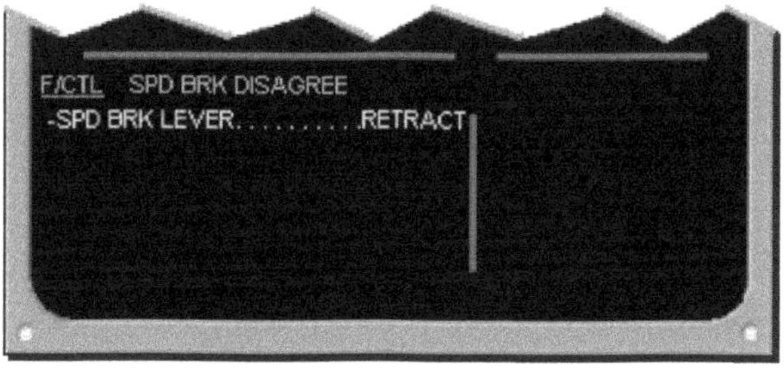

Each spoiler is fitted with a single actuator supplied by one of the hydraulic sources. In case of loss of hydraulic supply, an affected spoiler remains at its current deflection unless pushed down by aerodynamic forces. Spoiler control is shared by the 3 SEC computers. When a spoiler has failed on one wing, the symmetric one is automatically inhibited. As a general rule, spoilers are automatically retracted when faulty, inhibited or not electrically controlled. Each spoiler is fitted with a single actuator supplied by one hydraulic source. In case of loss of hydraulic supply, an affected spoiler remains at its current deflection unless pushed down by aerodynamic forces.

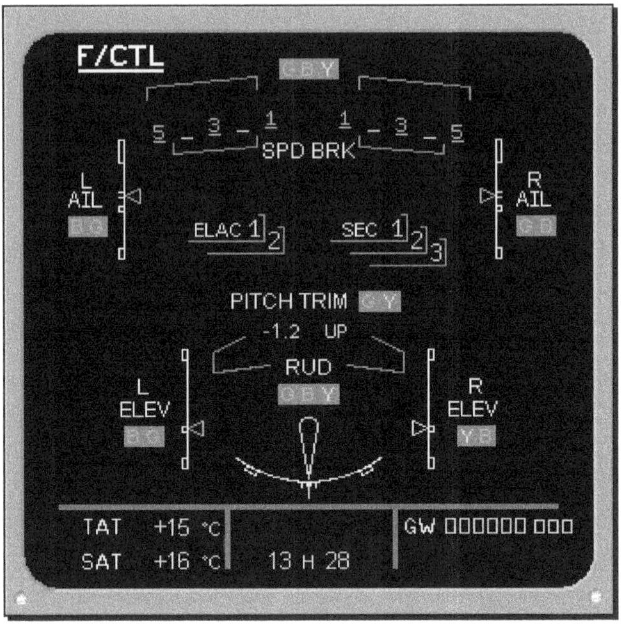

www.ingramcontent.com/pod-product-compliance
Lightning Source LLC
LaVergne TN
LVHW071742071126
851857LV00045B/2067